THE **COMPLETE GUIDE** TO

and
College
tial

Libr

POSTURAL
TRAINING

To my wife Suzanne, who continues to motivate and support me in everything I do; and to my daughter Maya India, who inspires me to move and play in ways that only she can imagine.

THE COMPLETE GUIDE TO

Kesh Patel

POSTURAL TRAINING

A & C Black • London

First published 2008
A & C Black Publishers Ltd
38 Soho Square
London W1D 3HB
www.acblack.com

ISBN 978 0 7136 8693 7

A CIP catalogue record for this book is available from the British Library.

Cover image © istockphoto.com
Inside photography © Grant Pritchard, except pp. 15, 16, 17, 18, 140 and 141 Kesh Patel
Illustrations © Jeff Edwards

This book is produced using paper that is made from wood grown in managed, sustainable forests. It is natural, renewable and recyclable. The logging and manufacturing processes conform to the environmental regulations of the country of origin.

Typeset in Baskerville by Palimpsest Book Production Ltd, Grangemouth, Stirlingshire

Printed and bound in China by South China Printing Co

CONTENTS

PREFACE AND ACKNOWLEDGEMENTS

The driving force behind this book didn't arise from a desire to 'correct' posture, but rather to re-educate it. I strongly believe in the process of re-education within this context, because we already know how to have good posture: we learnt the fundamentals of it during the first two years of life. This is an important observation, because in no form of animal life, including humans, is posture an inherited habit. It's an immediately occurring interaction between the individual organism and the force of gravity: an interaction that is highly adaptive. Although posture may be predisposed by the hereditary characteristics of our locomotor systems and the habits of our immediate family, it is entirely governed by how we react to the fixed laws of gravity. The close relationship between our posture and the locomotor system also strongly suggests that posture isn't just a static concept. For this reason, and throughout this book, posture will be regarded as an experience that applies to both static and dynamic conditions.

The interesting point is that we built most of the foundations for posture without much personal experience, a process that undoubtedly involved a huge amount of trial and error, and body awareness. Apart from the people around us, who were already upright, and speaking a language we couldn't yet understand, we had to rely on something else to teach us about posture. As small children, we probably have very little memory of how we learnt to sit, crawl, stand and walk unaided; as adults, we now take it for granted that we can stand upright and move without too much conscious thought. Yet we rarely ask ourselves how we came to assume an upright stance with so little formal tuition.

It's the answer to this question that has fuelled my interest in posture for many years, and has helped me to discover a number of useful approaches to re-educate posture. Of course, when learning anything for the first time, we often don't get the chance to appreciate the subtleties, nuances and finer points of the subject, until much later. The knowledge, reasoning and techniques contained within this book hopefully reflect such distinctions through many years of study, practice and personal experience.

To help me on my way, I had the continual support from both classical and contemporary schools of thought and, to this effect, I consider nothing in this book to be new. The philosophies of F. Matthias Alexander, Moshe Feldenkrais and Thomas Hanna in the field of somatic education are reflected throughout; the application of bio-physics, as conceptualised and realised through the work of Richard Buckminster Fuller, Ida Rolf and Thomas Myers is also acknowledged; the theories and ideas contained within evolutionary biology and anthropology, including the work of Frans de Waal, Robert Sapolsky and Frank Forencich are echoed throughout the text; the forward thinking concepts and ideas surrounding systems theory, cybernetics and neuro-linguistics, as developed by Gregory Bateson, Alfred North Whitehead, and Richard Bandler, also have their place; the highly effective techniques of therapists such as Vladimir Janda, Florence Kendall, Leon Chaitow, and Chris Norris are littered throughout; and lastly, and by no means least, the ideas and principles that govern human movement and performance today are also recognised, as illustrated so well in the methods of past and present physical educators such as Georges Hébert, Alvaro Romano, JC Santana, Paul Chek, Gray Cook, and Mel Siff.

The journey that I began many years ago, as a physical educator, continues to this day stronger than ever. It continues in my work, it's apparent when I teach, it's evident when I play – it's integrated into every aspect of my life. Now, it is strongly reflected in this book. The work that goes into writing any book, physically, mentally and emotionally, is phenomenal, and it would be impossible to adequately acknowledge and thank the many people who contributed to its evolution; however, I shall do my best to include everyone.

I am especially indebted to my wife Suzanne Patel, having been the lucky recipient of her wise editorial comments, relentless motivation and constant support; her efforts have helped me get my writing into shape.

A special acknowledgement goes to the team at A&C Black, London, and in particular, to Lucy Beevor, who took the time to go where no editor has gone before – to one of my Postural Re-education seminars.

As a practitioner, a heartfelt appreciation goes to the countless number of people who have sought my help for rehabilitation, conditioning and performance; by working with you, I continue to better myself.

As a teacher, I thank each and every student I've taught – the old dog has learned a few of your new tricks. Deserving a special mention are the trainers and therapists that I have worked with, who continue to keep me on my toes; especially to Alain Michelotti for referring such interesting patients to my practice, and whose support and wisdom has always been refreshing.

I would also like to thank all the physical educators of this world, past and present, who believed in what they were doing and continued to push the boundaries of philosophy and practice: you know who you are. Your actions speak for themselves and I feel privileged to be able to include your experiences and ideas in this book. You have taught me that when you're heading in the right direction, you know it's right because you're sure enough to be unsure, but never unsure enough to not do it.

The journey that is education is often shaped by the attitudes of the educators, and not necessarily by the knowledge they impart. On rare occasions, we become so enlightened by their attitudes, that we may find ourselves completely immersed in the moment, as the wealth of their knowledge and experience pervades our own thoughts and ideas. I would like to acknowledge three such educators who continue to inspire me in this way. To Frank Forencich, Richard Bandler and the late Moshe Feldenkrais – you have all taught me that in order to change the way we move, think and feel, we must first change the image of ourselves that we carry within us.

PART **ONE**

THE PRINCIPLES OF
POSTURAL TRAINING

INTRODUCTION

This chapter will introduce you to some of the key concepts surrounding the subject of posture in terms of assessment and re-education. It aims to provide practitioners of manual therapy, exercise and movement specialists, patients, exercisers, and anyone interested in human movement, with a number of useful and effective tools that will improve postural control, both structurally and functionally. Hopefully, this book will not overcomplicate the process of assessment and re-education – not all the information here will be necessary for your needs. It has, however, been put together in such a way that you can develop your own re-education programmes based on a firm foundation of classical and contemporary schools of thought.

Defining posture

It is difficult to define posture as it is interpreted differently by different people. The military officer, the teenager, the model, the dancer, the personal trainer, the actor, the anthropologist, the martial artist, the sculptor – these people and many more will all have their own ideas about what posture means to them. With this in mind, it is difficult to create a postural 'norm', a set of standards against which posture can be measured. Were such a norm to exist, all the above people would have a common point of reference. However, over a period of time, these postures become each person's own norm, and begin to feel so right and balanced that movement towards any other posture may feel unnatural. In this way,

temporary attitudes become long-term habits as the body moulds itself into fixed patterns that can influence current and future performance and functioning.

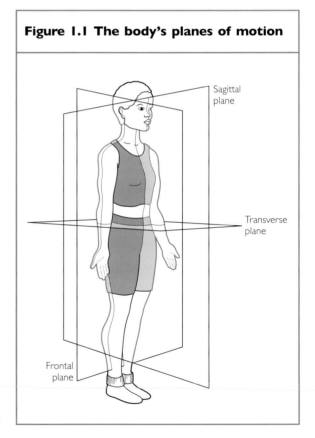

Figure 1.1 The body's planes of motion

Sagittal plane

Transverse plane

Frontal plane

Posture is often thought of in geometrical or Cartesian terms, where an ideal posture consists simply of the stacking of one joint on top of the other in relation to three planes of motion and axes of rotation – a configuration reinforced by good muscle balance around the joints (see fig 1.1). While this is true to some extent, it fails to recognise the way in which we achieve this configuration and how we are able to maintain it during movement. To look at these points, we must delve deeper and consider posture from another point of view – efficiency of use.

Although there are an infinite number of possibilities in the configuration of limbs and their mechanical use for any given task, inherent in every task is a way of organising and using the body with the least wear and tear and the most efficient functioning. For example, sitting down on a chair can be performed in many ways; however, only a few of these will require the least use of energy and effort. Much of the time, we become accustomed to using excessive amounts of energy when moving; indeed, many of us exhibit unnecessary amounts of muscular tension even when sitting or standing. This raises the question not of what's right or wrong, but rather what works? What positions or movements allow for optimal efficiency? The answers to these questions presuppose that posture is not necessarily defined by static positioning alone, but more importantly by the way we organise our bodies as we move. When such an attitude is adopted by practitioners, this type of questioning can effectively move patients towards a postural re-education programme that is motivated by a process of self-exploration and body awareness: a situation that is almost childlike in nature, and one that allows the body to re-learn and self-regulate its own posture.

The development of posture

The human body is a homeostatic organism. It is constantly adjusting, modifying and learning in response to internal and external stimuli: a process of self-regulation by continual feedback. These stimuli imprint themselves physically and mentally on the body in the way of learning – some desirable, some not so desirable. In fact, much of this learning begins before we are even born. Our ability to move begins in utero, we perform slow controlled movements, as well as rapid ones, almost as if we are stretching and strengthening in some way – a process known as pandiculation. These innate movement patterns appear to be directed by the precisely controlled and progressive expression (and inhibition) of important reflex pathways. When we are born, these patterns continue to help our learning as we lift and turn our heads, as we learn how to crawl, roll and sit, and eventually as we learn how to stand and walk. The systematic emergence of these movement patterns are important developmental milestones that can provide us with knowledge of how postural control evolved from birth. Much can be learnt from infantile movement patterns and the realisation that all children learn how to control posture in a very short space of time. This process implies a rapid learning curve, one that can be applied to adult re-education.

Further understanding of posture may require explorations into our ancestral heritage, and investigations into how and why human bipedalism has evolved and adapted over millions of years. Considering the extent of musculoskeletal dysfunction and movement impairment in today's society, the role of posture in this process, and an appreciation of how we used to move may provide clues to the restoration of optimal posture and function.

Lastly, postural re-education strategies should, perhaps, incorporate and integrate concepts of animal behaviour, in particular, those of other primates such as the great apes. It is often forgotten that humans are primates and as such we share much of our framework and potential motor behaviour with other apes. The overall aim is subjectively to model the origin of human movement from a diverse range of study options, and use this information to design effective postural re-education programmes in adults. A number of these principles are reflected in the postural lesson plans in chapters 7–10.

Control of posture

As adults, we often take our ability to sit, stand and move for granted – that is until we lose the ability to do so effortlessly. Unless this education is continued in the same way as it started, we can easily lose the ability to organise our limbs effectively and move efficiently. We may also lose the ability to notice the very stimuli that helped us to organise ourselves in the first instance. This inherent ability is known as feedback and feed-forward, and is largely governed by our proprioceptive and exteroceptive systems, which provide postural adjustments accordingly. The proprioceptive system senses changes in body position, posture and movement via specialist receptors that feed back information to the central nervous system. The exteroceptive system senses changes in our immediate environment through the senses of vision, hearing, touch, smell and taste. The basic principles of proprioception and how these intricate mechanisms help us to control stability and balance by compensatory and anticipatory postural adjustments are discussed in chapter 2. Using this knowledge, these same mechanisms of postural control can be enhanced further through appropriate exercise and movement. In this way, the nervous system can once again learn to interact continually with the environment, so that we can organise our bodies quickly and efficiently.

Evaluating posture

Optimal posture is more efficient than any other type of posture: the proprioceptive mechanisms are beautifully integrated to bring the body into a configuration where all weight-bearing articulations are subjected to minimal compression and stress. As we attempt to 'measure' posture, we may initially observe such a configuration statically, through observations of joint alignment. We may also look at how these characteristics affect our ability to function dynamically, by assessing movement potential, balance and coordination.

The assessment of static posture is usually conducted against a gravity line, which provides a point of reference for observation of joint alignment. These observations can then be correlated with those of seated or lying assessments, and later confirmed with specific muscle function testing. This provides an overall picture of the way the body organises itself statically in relation to gravity, and is also known as structural efficiency. Variations away from the gravity line can be interpreted within the context of existing pain and impairment, or can be used as indicators of potential dysfunction. Some common postural variations are discussed in chapter 4.

The assessment of dynamic posture usually involves the performance of key movement patterns that challenge a number of important postural mechanisms, such as coordination, balance, economy of effort, and reversibility. This provides an overall picture of how the body organises itself in preparation for and during movement, and is also known as functional efficiency. A number of important movement assessments are explored in chapter 4, with discussions of why they are vital to postural health.

Putting this altogether, we now have a different model for posture: one that focuses on body efficiency or how the body is used for any given task. From an evaluation perspective, this doesn't stop at observation of joint alignment. In fact, a more suitable term might be body configuration, where configuration relates to how parts of the body are arranged and interconnected, so that the body functions correctly.

A model for re-education

Many of the techniques presented in this book are synergistic in the sense that they are more effective when they are performed as a sequence, rather than when they are performed in isolation. Each movement sequence aims to meet a specific postural objective, and movements are organised in such a way to maximise learning and change. To help this learning process, a number of therapeutic practices are used in a specific way that allows accumulative change, in that each method effectively builds on the previous method in the sequence. Many of these practices are discussed in chapter 6, and include positional release and contract-relax techniques, which can then be effectively programmed alongside integration techniques, to build postural lesson plans.

The objective of each lesson plan is not to correct a particular postural dysfunction, but to allow you to focus on how you are using a particular body part in relation to other body parts. The result is two-fold:

1. Effective resolution of pain and dysfunction, as the nervous system recalibrates itself to detect postural disturbances at a finer level.
2. Increased efficiency of movement and economy of effort, as the body learns to organise itself differently in relation to itself and the environment.

Such a change in self-perception, whether the objective is to eradicate pain, or to prevent pain and maintain good health, is an important result of postural re-education. In the majority of cases, this process involves improving body awareness, an approach that is heavily reliant on effective 're-wiring' of the body's sensation of position and movement. As previously mentioned, a number of techniques can be used to achieve this, which can be self-administered or practitioner-assisted.

These techniques have been organised into 20 lesson plans, thematically organised around the achievement of common postural objectives, and can be found in chapters 7–10. Each lesson has been developed from a combination of knowledge and experience. Some focus on the frequent repetition of carefully selected exercises, performed with control and precision; others favour a more cognitive approach whereby neuromuscular pathways can be rebuilt by influencing your thought processes, with a strong emphasis on techniques such as dynamic imagery, where visualisations of favourable body positions and postures are used during exercise and movement. Some solely focus on restoring somato-sensory awareness using developmental movement patterns. In all cases, an emphasis is placed on self-exploration, heuristic movement, and continual feedback, where there is no concept of right or wrong – only what works.

What to expect

This book presents effective strategies for improving posture and subsequent functioning of the human body. It is aimed at everyone seeking to make positive changes to their own structure and function, as well as those who have a clinical interest in facilitating this process. Rooted deep within these strategies is the assumption that the body doesn't have to be taught how to do this; instead, it needs to find a way to re-learn and to continue the education that it has already received early on in life. As you read the text and use the techniques, many of these concepts will come to light, and you will invariably find yourself integrating them into your daily life, making lasting changes to your own posture.

CONTROL OF POSTURE

Introduction

The control of human posture is multi-faceted, involving numerous and complex mechanisms. These 'control systems' not only keep us upright when standing and during movement, but also to maintain equilibrium and spatial awareness in all manner of body positions. For the human body to effect these changes, at least four things are needed:

1. We require structures that sense incoming information from the internal and external environment.
2. We need to receive this information accurately through the central nervous system.
3. We need to adjust our posture in accordance with the received information.
4. We need a mechanism that allows the body to compare these adjustments to an existing standard. The result is a closed loop control system that maintains a balanced and restful state, or postural homeostasis (*see* fig. 2.1).

This level of homeostatic control, which involves precise communication between the nervous system and musculoskeletal system, highlights the self-regulating characteristics of the human body. This process involves a number of finely-tuned mechanisms that work together to control posture centrally. These include:

- Proprioception and exteroception
- Reflex movement
- Balance and stability
- Compensatory/anticipatory adjustments

As human beings we take for granted that we have a good degree of control over our unique upright posture. To some extent, this is justified; most of the time, we are able to perform daily activities without placing conscious attention on our posture. However, this way of thinking can give us a false sense of security, because when musculoskeletal problems that affect posture arise, we are unable to regain control and maintain equilibrium once again.

This chapter aims to discuss the foundations of postural control as they relate to proprioception and exteroception, reflex movement, balance and stability, and postural adjustments. While comprehensive details of these mechanisms will not be explored, particular references will be made to their application during postural evaluation and re-education.

Proprioception and exteroception

Proprioception refers to the kinaesthetic awareness of body posture, position, and movement, both as a whole, as well as in relation to neighbouring parts of the body. It is a sense that provides feedback solely on the status of the body internally. This feedback comes from the cumulative neurological input of a number of 'transmitters' (proprioceptors) that sense changes in the internal body environment and subsequently send this information to the central nervous system for processing. The majority of the proprioceptors responsible for postural control are located within the paraspinal

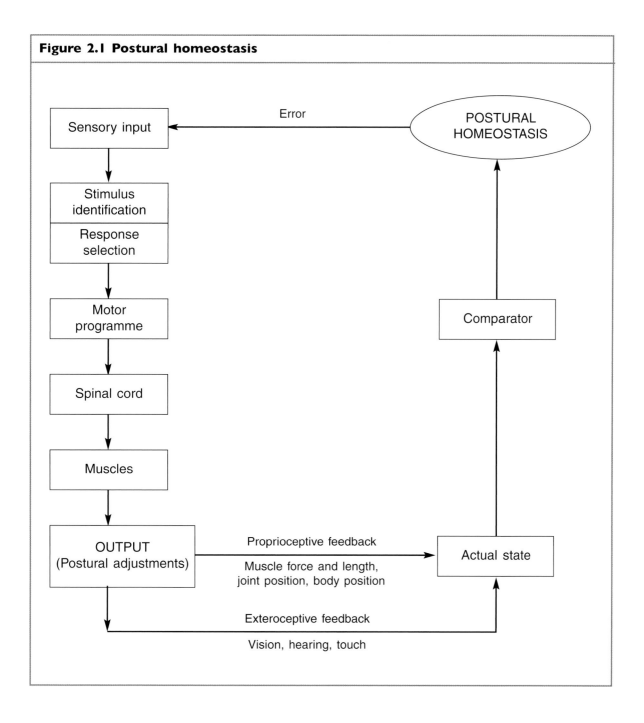

Figure 2.1 Postural homeostasis

and suboccipital muscles, in the labyrinths of the inner ear, and in the muscles and joints of the lower extremity.

Exteroception refers to the five senses of sight, hearing, touch, taste and smell, that feedback information about the outside world. Out of all of these, sight, hearing and touch are the most important for control of posture. Vision and hearing are intimately related to the vestibular system, and touch (especially on the soles of the feet) is an important element of postural control. Exteroceptive feedback is continually collaborated with proprioceptive feedback to integrate information about posture and movement.

Reflex movement

A reflex movement is a rapid and specific pattern of response that occurs without volition and without the need for cerebral input. Automatic reflex movements accompany all normal voluntary motion and, indeed, very few muscles in most movement patterns are under conscious control. This last point illustrates a potential direction in postural re-education – one that encourages natural movements that focus on integration of body segments, rather than isolating individual muscles.

Reflexes may be simply classified into two groups: proprioceptive and exteroceptive. Proprioceptive reflexes are generally those that take place in response to changes in motion and position of the body. This mainly occurs by stimulation of receptors found in skeletal muscle, tendons, joints and labyrinths of the inner ear. Important examples include the stretch reflex, the tendon reflex, and the labyrinth and neck reflexes. Exteroceptive reflexes are those that take place in response to changes in the external environment. Examples include the rapid flexion response of the body upon hearing

a sudden loud noise, rapid closure of the eyes when a foreign body threatens to hit the eyeball, or scratching of the skin when irritated.

Balance and stability

Any body at rest is considered to be in equilibrium, with all forces acting on it being balanced; however, not all bodies at rest are equally stable. A stable equilibrium occurs when an object's centre of gravity is in the lowest possible position, so that any effort to disturb it would require its centre of gravity to be raised. A neutral equilibrium exists when an object's centre of gravity is neither raised nor lowered when it is disturbed (for example, a ball). The human body has a different type of equilibrium known as unstable equilibrium, where the centre of gravity is as high as possible, and therefore drops towards a lower point when the body is tilted. At first glance, this seemingly precarious position of balance appears to offer no advantage whatsoever. However, with more insight, it is possible to see the benefits of such a design.

Moment of inertia

Compared to other animals, the human body requires very little energy to set it in motion; it is said to have a low moment of inertia around the line of gravity. The low moment of inertia is a direct result of the vertical stacking of the pelvis, trunk and head over a small base of support, ultimately placing the centre of gravity in the highest possible position. In this position, the amount of muscle contraction required to maintain the body is minimal. This also suggests that the human body is poorly designed for standing still, and is in fact more suitably designed for movement – an action that requires very little effort to initiate.

Developmental considerations

Developmentally, movement is achieved much more easily than immobility. Toddlers begin their journey towards unsupported standing by taking a series of forward-falling steps, long before they are able to stand unsupported and motionless. In fact, years can pass before a child is able to stand on one leg, and many adults can never do it, in spite of being able to perform many different movements.

The human body is therefore always ready for spontaneous movement in any direction, with almost no expenditure of energy; what little energy is expended comes from its potential energy, which is restored after the movement ceases. Although other animals may exhibit faster movement in one particular direction, humans have the potential for all-round freedom of movement, a potential that is often compromised due to musculoskeletal dysfunction and pain.

A question of efficiency

When considering standing, we can conclude that any deviations from the position described above may be thought as inefficient. It is important to note that inefficiency does not mean failure, nor does it imply that something is wrong; in this context, it simply suggests an energy-costly movement away from a position of optimal efficiency. In reality, many individuals are able to function well enough with seemingly poor efficiency of movement, and very little pain. However, the question remains as to how long this inefficiency is sustainable without compromising structural and functional health.

Factors affecting stability

Understanding the factors that affect the stability of the body's state of equilibrium is essential as a means for re-educating posture. This begins with the simple yet often over-looked appreciation that our ability to maintain balance over our feet is one of the basic motor skills. From an early age, we learn how to balance our whole body, moving from balance over four points to balance over two, in a relatively short period of time. As adults, we often take it for granted that we can balance effectively on two feet, but such a task is extremely challenging and requires optimal motor control.

There are six important factors that affect stability:
1. Base of support.
2. The line of gravity in relation to the base of support.
3. The height of the centre of gravity.
4. Segmental alignment.
5. Friction.
6. Ocular and vestibular mechanisms.

Base of support

The base of support is by far the most important factor that affects stability. At birth, the angle of inclination at the femoral neck is approximately 140–150°, which gradually decreases to 125° over the first few years. Such a large angle widens our base of support, allowing the centre of gravity to remain over it with ease. This large base of support may also help to offset the slightly higher centre of gravity due to a relatively larger head size compared to the rest of the body. It is important to understand that the base of support includes not only the part(s) of the body in contact with the ground, but also the space between them; if the contact points are separated, the base widens, and equilibrium is improved.

A factor often overlooked is how the shape of the base of support also affects stability. Because the human body functions in a multi-planar environment, it is important to learn how to manipulate the shape of the support base to maximise stability in relation to the direction of applied force. Unfortunately, this is a motor skill that is often forgotten in adults, but it can be re-educated with appropriate training. For example, when standing on a train, there may be a tendency to be thrown backwards if you are standing in the direction of the line of force. To resist the external force and remain stable, you should widen the base of support in the direction of the oncoming force (i.e. stand sideways with a wide stance). The sole purpose of this adjustment is to keep the centre of gravity over the base of support. When faced with an unpredictable force, it is best to adopt an oblique or bracing stance, similar to that of a boxer or martial artist, a position that is often adopted automatically in such a situation.

The line of gravity in relation to the base of support

The body will only remain stable as long as the line of gravity falls within the base of support. In upright posture, this means that for maximum stability the centre of gravity (situated at the level of S1, 8–13cm below the belly button) should ideally be in approximately the same vertical plane as the arches of the feet. If the line of gravity deviates towards the boundaries of the base of support, then stability may be compromised.

If we apply this concept to posture, we can see that efficiency will be greater when the major joint axes are aligned above one another, and close to the gravity line. In reality however, this 'ideal' relationship is often compromised, and joint misalignment occurs – resulting in body segments that have deviated away from the gravity line. In these situations, the body has a tendency to rely more on muscular support, and this, if excessive, may expend large amounts of energy, as well as contribute to joint and muscle pain over periods of time.

Deviation of the line of gravity over the base of support is often seen when carrying a heavy load in one hand. In this situation, a compensatory movement of the opposite arm outwards, or leaning over to the opposite side, allows you to counterbalance the load effectively. This is a useful action, although it may produce significant deviations in joint alignment and undesirable muscle forces, as part of a repetitive movement pattern.

The height of the centre of gravity

As previously mentioned, the body's centre of gravity while standing is located at the level of S1. However, when the arms are raised overhead, or when a load is carried above this level, the centre of gravity assumes a higher position, thus making it difficult to maintain equilibrium. Conversely, lowering the centre of gravity will increase the stability of the body. This is easily seen, for example, in a crouched position with your hands out to your sides on the floor; the centre of gravity is now positioned over a wider base of support.

It is important to note that while a lower centre of gravity increases stability, it increases the energy cost of movement. The naturally high centre of gravity of upright posture means a low moment of inertia, making the body ideally suited for movement. The height of the centre of gravity can be manipulated during movement to alter stability, providing a useful means of improving postural control.

Segmental alignment

The alignment of body segments or links is widely used to evaluate static and dynamic posture. When the centre of gravity of all weight-bearing segments lie in a vertical line positioned over the base of support, posture is not only stable and bio-mechanically sound, (in terms of minimal strain to the joints and muscles), but also aesthetically pleasing. The joints and segments cannot slide off one another, and they are bound by certain constraints – fascia, ligaments, tendons and capsular structures. However, some movement is allowed. This freedom of movement can lend itself to 'zigzag' misalignments, where one segment gets out of line and results in a compensatory misalignment of another segment. At every point that this occurs, there are potentially uneven tension patterns in the soft tissue around the segments or joints. Initially, this may lead to premature fatigue of the associated structure, eventually leading to strain and/or pain. A common example of this occurs when your head is carried too far forward of the axis of your shoulder joint. This results in unnecessary tension in the neck, and strain of the upper back muscles, as the thoracic spine increases its curvature to compensate for the forwardly displaced centre of gravity.

When an external load is added to the body, such as a shopping bag or a baby, it should be regarded as an additional segment added to the body. Stability will be affected by the height of the load, its mass, and also its position in relation to the body's centre of gravity. For example, a heavy load that is held at arms' length will make the body more unstable than the same load held close to the body. In this case, stability can be maintained by increased contraction of the associated musculature. All too often, many people fail to organise their bodies effectively along the line of gravity, or to recruit the appropriate muscles to enhance stability; a situation that occurs either through lack of use, decreased awareness or incorrect teaching. The result is compensatory muscle action and possible strain or injury.

When re-educating posture, it's important not only to consider what demands are placed on the body, but also how they are affecting the body in relation to the centre of gravity and segmental alignment. Because overcoming gravity occupies a significant portion of the activity of our nervous system, re-education may often begin when lying down in order to free up resources for learning, as muscles begin to relax further and reduce their pull against gravity. This may then be followed by standing exercises which can be loaded appropriately. The overall aim is to teach the patient how to maintain postural homeostasis, increase efficiency of movement, and remain pain-free during daily living.

Friction

We live in a world where most of our stability and movement relies on adequate frictional forces; these forces initially act where the foot contacts the ground. When the supporting surface has insufficient friction, the body can draw upon both intrinsic and extrinsic strategies in an effort to maintain equilibrium. Extrinsic factors include modifications to footwear such as the use of rubber soles or spikes. The most obvious spontaneous intrinsic strategy is for you to lower your centre of gravity by widening the base of support, or crouching. In activities or sports where there is prolonged exposure to inadequate friction on the support surface, righting and tilting reflexes as well as specific muscle co-contraction patterns will be enhanced, which will maintain stability and balance. When designing postural re-education programmes, the exact environmental conditions of inadequate friction do not necessarily have to be simulated; instead stimulation of associated reflex pathways and muscle recruitment patterns may be achieved using strategies such as stability balls, balance boards and slide boards. Once stimulated, these patterns can then be re-integrated into original movement patterns with greater efficiency.

Ocular and vestibular mechanisms

The maintenance of upright posture and balance has an intimate relationship with the ocular and vestibular system. Appropriate positioning of your head allows your eyes not only to see with less effort, but also to sense accurately changes in the surrounding environment by your peripheral vision. A balanced head position will also allow for optimal circulation of blood to the eyes. During movement, your eyes help to maintain equilibrium by focusing on stationary points at the level of the horizon. This mechanism is highly efficient in individuals who perform acts of balance or rotation, such as gymnasts and dancers; the focus on a point further from the area of immediate danger can help to control balance. The level of equilibrium afforded by the ocular system is easily seen when attempting to perform even the simplest of movements with your eyes closed. Unfortunately, modern day living habits often put us into situations where our eyes are aimed downwards to read, write or perform certain tasks. As your eyes move down, so your head follows, and in time this position becomes habitual. The consequence of this is that when your eyes are raised to the horizon, your head is pulled back around the atlanto-occipital joint, resulting in unnecessary tension patterns and less than optimal alignment of the cervical vertebrae.

The vestibular apparatus of the inner ear registers variations in pressure from outside and inside the body, and feeds back this information to the central nervous system making us aware of our movements and where we are spatially. This is achieved through the labyrinths, which contain the semicircular canals – essentially a system of spirit levels. Each canal is filled with a fluid and has hairs attached to the cavity walls. The hairs move as the fluid moves inside the canal in response to any angular acceleration, and feeds back information about body position.

With the growing acceptance of the importance of both these systems in the maintenance of equilibrium and balance, a number of highly effective vestibular exercises can be used and integrated effectively with more traditional approaches.

Postural adjustments

Our upright posture is inherently unstable with the problem of balancing the body, which has a high centre of gravity, on a narrow base of support. Despite this apparent instability, the muscle effort required to maintain upright posture is minimal. When voluntary movements are performed while standing, the redistribution of body mass and inertial forces jeopardises our ability to remain balanced and stable, and causes a number of postural adjustments.

Two main situations exist where postural adjustments take place in fixed patterns with several groups of muscles responding in a stereotypical way. The first example is the correction of excessive postural sway; the second is anticipatory and compensatory postural adjustments.

Postural sway

When standing still, there is normally a small amount of movement in the neck and trunk, which produces a pendulum-like motion in the centre of gravity, in the frontal and sagittal planes. As the centre of gravity shifts in this way, there is a concomitant shift in the centre of pressure – the point at which ground reaction force is applied through the ground-foot interface. This rhythmic pattern of movement is known as postural sway, and is usually restrained to no more than 10mm of movement. Increased magnitude and velocity of sway may result in problems with stability that are associated with age, fatigue and environmental influences, as well as pathological conditions, such as vestibular dysfunction.

There are two strategies for correcting postural sway during standing (Horak and Nashner, 1986): the ankle strategy and hip strategy.

The ankle strategy

When standing on a base of support that is larger than your foot, sway is corrected by activation of either the anterior or posterior muscles, starting at the ankle and radiating successively up to the thigh and trunk muscles. For example, forward postural sway will activate the plantar flexors, immediately followed by the hamstrings, and then immediately followed by the para-spinal muscles, with delays of about 20 milliseconds between each. During backwards swaying, the tibialis anterior, quadriceps and abdominal muscles are activated in sequence. In both cases, activity begins in the ankle joint muscles and then radiates in sequence to thigh and then trunk muscles. This activation pattern exerts torque around the ankle joints. It is known as the ankle strategy because it restores equilibrium by moving the body primarily around the ankle joints.

The hip strategy

When standing on a base of support that is smaller than your foot, changes in the angle of your ankle do not contribute to stability and balance.

In this situation, an antagonistic strategy is used whereby the trunk muscles are recruited, immediately followed by the thigh muscles. For example, during forward sway, the abdominal (proximal) muscles are recruited immediately followed by the quadriceps (distal). This activation pattern produces a compensatory horizontal shear force against the support surface but little, if any, ankle torque. This pattern is known as the hip strategy, because the resulting motion is focused primarily around the hip joints.

It is unlikely that both of these strategies are caused by stretch reflexes alone – rather they are centrally coordinated patterns of movement. It's important to note that in the absence of somato-sensory input, the hip strategy is preferred; in the absence of vestibular input, the ankle strategy is used.

Anticipatory postural adjustments

Central control of posture during complex movement patterns is achieved through anticipatory and compensatory postural adjustments, which serve to maintain a frame of reference in conditions of possible external and internal disturbances (Massion, 1992). This frame may refer to the position of a body segment, to the whole body, or to general equilibrium, i.e. keeping the projection of the body's centre of gravity within the base of support. Any voluntary movement, especially a fast one, induces a postural disturbance and also shifts the centre of gravity. Therefore, voluntary movements involve disturbances that may be predicted to a certain degree by the central nervous system, which in turn adjusts the activity of postural muscles both before the actual disturbance and in response to it.

There are clear differences in the function and control of these two groups of associated changes in activity of postural muscles. In essence, the central nervous system tries to predict postural disturbances associated with a planned movement and minimise them with anticipatory postural adjustments (APAs). Compensatory postural adjustments (CPAs) deal with actual disturbances of balance that occur because of suboptimal efficacy of the anticipatory components. CPAs usually involve additional and altered muscle recruitment patterns as the body attempts to restore stability and balance, and to maintain a frame of reference.

Postural programmes can be designed to create situations in which postural disturbances take place to varying degrees. In this way, you can begin to fine tune your adjustment mechanisms, as they become increasingly aware of the changes in your body position. Further progressions should also raise awareness of the muscles of your upper limbs, and their role in helping to restore equilibrium. In this way, the central nervous system can learn to adjust the activity of postural muscles both before the actual disturbance and in response to it.

ASSESSMENT OF STRUCTURE

Introduction

Maintaining your balance under gravity by accurate alignment of the body's joints is essential to keep postural homeostasis. By observing joint alignment, we can draw reasonable conclusions about the structural integrity of the body as it relates to the way different body segments connect to each other and maintain position under gravity at any given time. The assessment of structure by way of observing joint alignment is an ongoing one and it is common for it to begin in the standing position against a gravity (plumb) line. Initially, the practitioner should focus their attention on any deviations away from the gravity line, as well as the extent of these deviations, and whether they are causing pain or discomfort when standing. Human beings come in all shapes and sizes, and a posture that 'fits' one person may cause pain in another. Therefore, if there is no pain where there is obvious deviation from the gravity line, it is important to make a note of it, without inferring too much at this point. It is quite possible for such deviations to come to light during functional assessment.

When viewing an individual against the gravity line, it is important to note that deviations in the sagittal and frontal planes may cause, as well as be caused by, changes in muscle strength and length. The reason for this is not important at this stage; instead, the focus should be on using these observations to design postural strategies that help to restore optimal alignment (in relation to gravity), and to enhance postural control.

Further information about musculoskeletal changes may also be gained through muscle testing, and joint and ligament assessment: an area that is beyond the scope of this book. Such tests may help to confirm changes in soft tissue structure and function, and may be an area for further treatment or referral. In chronic cases, where individuals have had postural deviations over months and even years, there may be morphological changes in the soft tissues – for example muscle hypertrophy/atrophy – and these are simple observations that should be duly noted.

The aim of this chapter is to explore the assessment of structure as it relates to joint alignment, and contains a number of important observations of standing, seated and lying positions. As the practitioner becomes more experienced with these observations, they will be able to streamline their techniques, resulting in a quick, effective and information-rich approach. Further study of muscle and joint testing techniques will also complement this approach. Many methods exist for the preparation and assessment of posture, and the practitioner should note that only an introduction to some of the more important techniques and observations are outlined in this chapter.

Preparation for assessment

The assessment of posture should begin by choosing the right tools for observation, as well as the appropriate set-up of the subject; the following guidelines will apply to both structural and functional assessments.

Evaluation tools

Many tools and gadgets exist that may help to evaluate posture. Most of them are objective in nature, such as plumb lines, inclinometers and goniometers, their success relying primarily on the parameters of a hypothetical 'ideal' for posture. While it is widely accepted that an optimal skeletal alignment exists (Kendal et al, 1993), the importance of relying on subjective measurements such as observation, should not be forgotten. In the clinical setting however, a combination of tools may provide the best approach in terms of quality and speed of assessment.

Observation

By far the most subjective and effective of tools is the simple observation of posture, yet it is often underused as an integrative and predictive tool. During assessment of static posture, the way in which an individual stands is often overlooked; an observation of increased postural sway may indicate poor postural control, or excessive muscle contraction, which may suggest poor alignment. These results may then be used to show potential limitations in efficiency. While it may be useful to assess movement patterns visually, such as bending or sitting, this assessment should also extend to observations of when such movements occur naturally, such as when an individual walks up the stairs or sits down on a chair unprompted.

Plumb line

The plumb line is the simplest of tools and is central to the assessment of posture. Consisting of a piece of string with a metal weight attached to the end, it is usually hung from the ceiling as a visual representation of the gravity line. This vertical line can then be viewed against body landmarks, such as the middle of the shoulder joint or knee joint, to ascertain how well aligned the body is with respect to gravity. Major variations away from the gravity plumb line in both the mid-frontal and mid-sagittal

planes may suggest potential areas of stress or strain, as muscles shorten or lengthen to maintain an upright body position.

Figure 3.1 A plumb line

The application of a plumb line may be extended to the use of a postural grid. This consists of a wall- or door-mounted screen, usually no larger than the size of a door, on which a grid is printed. The screen should be levelled horizontally using a spirit level. Alternatively, a grid can be printed directly onto a blank wall. The use of such a grid allows the practitioner to assess posture using not only a series of vertical plumb lines, but also horizontal ones, which allows them to observe bilateral variations in alignment. A grid is also useful for assessing alignment during movements, such as squatting.

Digital photography

The use of digital photography as a means of recording posture can provide additional evidence during assessment. Still photos are often useful as back-ups for written or hand-drawn records, and may even serve as a motivational tool for some people. With the popularity of digital photo-editing software, plumb lines can be added to photos, but make sure the person is standing on level ground when the photo is taken.

Figure 3.2 Against the plumb line grid

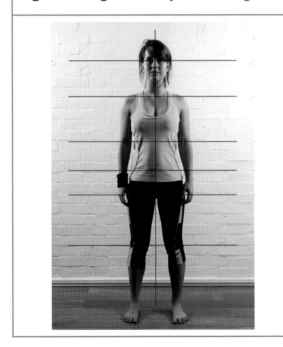

The use of video analysis or time-lapse photography is also a popular method of assessing movement patterns, from simple recordings against a postural grid, through to gait analysis and sports-specific movements. In all cases, the practitioner should seek the permission of the individual before using any type of photography or video analysis.

Foot plate

These are boards on which neutral footprints have been drawn to standardise the stance of the person being assessed. The footprints are drawn with a base of gait of about 8–15cm, and an angle of gait of about 10°. One of the reasons the footprints are standardised is to reveal fully the extent of any compensatory muscle adaptations such as shortening or lengthening. The inability to stand comfortably, or the presence of joint or muscle pain, may highlight areas requiring treatment.

Figure 3.3 A foot plate

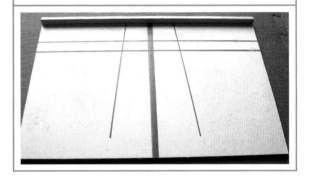

In addition to using the foot plate, the experienced practitioner should also observe the functional or habitual footprint, i.e. how the individual normally stands without instruction. As this is how most people will stand naturally during their daily life, it will give the practitioner an idea of how much variation exists from the standard stance, and provides them with a working reference point. It is important to remember that there may be considerable variation between the standard stance of different individuals; any variations should always be considered alongside known or potential structural and functional dysfunction.

Foot raises

Wooden foot raises are used during standing assessments to determine how much lift is needed to level the pelvis laterally, in the presence of a one-sided hip hike. Raises should be no bigger than the size of a large foot, and should be cut to 0.3cm; in this way, several blocks can be stacked if needed for a more precise correction.

Figure 3.4 Foot raises

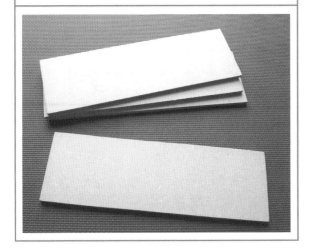

This technique is generally preferred to the use of leg length measurements, because of the differences in muscle recruitment when supine and standing (Kendal et al, 1993). Any observations should be backed up with muscle function testing, to confirm the presence of structural or functional causes of musculoskeletal dysfunction and/or movement impairment. In the case of structural variations, the results may suggest the need for foot orthotics, however this type of diagnosis should always be made by an appropriately qualified professional. Foot raises may also be used during functional assessments for joint re-alignment, to qualify the need for specific corrective exercises and movements.

Goniometers and inclinometers

A goniometer is a device used for the general measurement of joint angle and range of motion, whereas a posture inclinometer is an instrument used specifically to measure the degree of the spinal curvatures, as well as rib inclination and pelvic tilt. Such measurements are important as diagnostic tools or when designing specific rehabilitation programmes. They are also useful postural tools when a training programme fails to meet the objectives of rehabilitation and/or performance, and may suggest the need for alternative treatment methods. However, it is important to understand that when attempting to interpret posture as a whole these readings alone may be limiting; in such cases, the interpretation of results against a theoretical 'norm' should always be contextualised within the framework of structure and function. In this way, postural movements can be designed for the body as a whole.

Figure 3.5 A goniometer

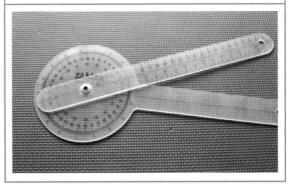

Flexible curved ruler

A curved ruler can be molded on to the entire length of the spine to provide an imprint of the spinal curvatures during static postural assessment, and is especially useful when the individual is clothed. It is also useful for new practitioners who are still mastering their visual and palpation skills. The imprint can also be used to provide visual feedback for the individual, as an aid to their understanding of their own posture. The use of this tool may be extended to imprinting the spine during end ranges of movement, such as the bottom of a squat, to highlight any major changes in spinal curvatures, the results of which can then be used as a means of re-education.

Figure 3.6 A flexible curved ruler

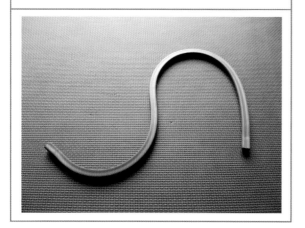

Wooden dowel

Like the flexible ruler, the wooden dowel is used as a reference point for spine alignment, by placing it vertically against the spine. Any deviations away from a reference position could indicate a number of things, including poor coordination or muscle imbalance. It may then be used as a re-education tool for optimal spine alignment during standing and movement.

Marker pen

A water-soluble marker pen can be used for marking bony landmarks such as the spinous processes, angles of the scapula, or patellae, during static and dynamic assessment. It can be very useful when looking at relative joint alignment during symmetrical and asymmetrical movements.

Set-up of the subject

Assessment of joint alignment usually begins with the person standing in bare feet. Because some footwear may temporarily correct postural distortions, it may also be useful to repeat observations with footwear on, to gauge how it affects posture from a functional perspective. The person should stand comfortably and evenly, as if planning to hold the position for several minutes. What is often observed when standing upright is that the stance often favours weight bearing on one side, with a corresponding hip-hike on the weight bearing leg. If the person assumes such a stance, it should be noted, and they should be instructed to balance their weight over both legs for the purpose of assessment. The outcome of supporting weight predominantly through one leg is still likely to be evident as a postural distortion during assessment.

Figure 3.7 Standing postural assessment – the set up

To help the person to find a relaxed and natural base and angle of gait when standing, the practitioner can ask them to march on the spot for a few moments so they can position their feet for balance and comfort. Observations should also be made with a neutral footprint, either using a customised foot plate or masking tape on the floor. A neutral footprint position is where the feet are positioned under the glenohumeral joints, with no more than 10° of lateral rotation (Chaitow & DeLany, 2002). Observations in this position will usually exaggerate any postural distortions, such as hip, shoulder and head tilt.

During the assessment preparation, the person should stand in front of or behind the plumb line, which should begin centrally between the heels, from the ground up. There should be ample space for the practitioner to move around the subject, and it is often useful at this point for the practitioner to stand back several feet and begin observations with an overall impression of joint alignment. This may help to expose any major postural distortions that might otherwise be hidden when viewed from close-up. Observations should be made anteriorly, posteriorly and laterally, to gauge the level of anterior–posterior and bilateral balance. Posterior observations will usually confirm anterior variations, as well as providing further information about the scapulae and spine. The point at which to start observing is down to personal preference but, to put the person at ease, it may be useful to begin in an anterior position from the ground up, as this is where the plumb line starts, followed by posterior and lateral observations.

Figure 3.8a Standing against a postural grid – anterior view

Figure 3.8b Standing against a postural grid – lateral view

The following information will provide the practitioner with an easy to follow and structured guide to assess static posture when standing, by posing a series of questions about specific body landmarks and soft tissue appearance. Although this approach will yield useful information about isolated parts of the body, it is always important to view these results in a context of how these parts integrate and affect each other. This holistic way of thinking is reflected in the design of the postural lesson plans that appear later in the book.

Anterior observations

The feet and ankles

Are the feet bearing equal weight?

This may be evident by the presence of excessive pronation through the foot bearing more weight; there may also be a lateral deviation of the person away from the central plumb line, as observed at the crown of the head. Excessive weight shifting can also result in lengthening and weakening of the same side hip abductors, causing a lateral shift in the hip (and pronation). This may be confirmed with abductor testing, and function can be restored with appropriate gluteal re-education.

Do the feet have normal arches without excessive pronation or supination?

Often, observation confirms the presence of pronation or supination but it may not be easy to quantify for the purpose of subsequent treatment. If either variation exists, it's important to ask the subject whether it is problematic (i.e. local tenderness/pain, impaired movement, medial knee pain etc); this will help to ascertain the extent of the functional variation. The presence of calluses or bunions on the inner or outer aspects of the foot will also highlight the degree of pronation or supination respectively.

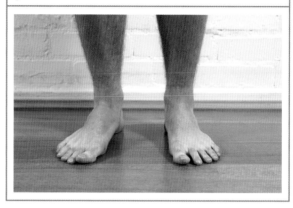

Figure 3.9 Arch support and ankle position

Long-term pronation may also contribute to medial knee pain, particularly if the lateral rotators of the hip are also weak. The potential muscle imbalances contributing to pronation include weakness in the tibialis posterior and tightness in the peroneus longus. Tightness and weakness of the medial aspect of the gastrocnemius may also be a contributing factor.

Structurally, the degree of pronation/supination may be objectively assessed in three ways:
1. By posteriorly observing the degree of bowing of the Achilles tendon, from a vertical position.
2. By observing the degree of deviation away from a vertical reference that is positioned at the outer aspect of the foot in line with the lateral malleolus.
3. By placing the foot into a neutral position, and observing the difference.

Improving control of the foot and ankle muscles can be very successful in restoring optimal alignment to the ankle; long-term postural control will also require re-integration of the foot and ankle into larger lower extremity movements such as squatting and lunging.

Are the feet in line with each other, with the heels approximately 8–15cm apart, and with out-toeing no more than 10°?

Having one foot forward of the other may indicate pelvic torsion or shortness in the same side hamstring, as the knee flexes and places the foot forward. In this case, there may also be more weight distributed on the opposite leg, potentially resulting in a hip hike on that side (hip joint adduction). If this is the case, the abductors may be weak.

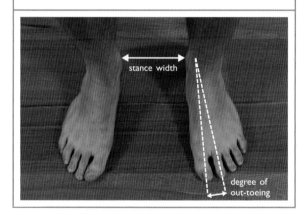

Figure 3.10 Observation of stance

stance width

degree of
out-toeing

An excessively wide stance may be associated with tightness in the hip abductors, and may also imply weakness in the hip adductors.

Excessive out-toeing when standing is associated with tightness in the lateral hip rotators; a small contribution to out-toeing may be made by lateral tibial torsion if the knees are slightly flexed. Out-toeing is commonly associated with pronation, particularly in the presence of a structural tibial torsion.

The knees

Are the kneecaps level horizontally?

This may be assessed by placing a finger on the top of each patella; un-level patellae may indicate a discrepancy in the structural or functional leg length.

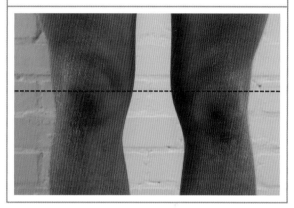

Figure 3.11 Position of the patellae

The dotted line passes through the centre of the kneecaps.

Excessive tension in the quadriceps may also pull the patellae upwards.

It is important to consider the compensatory pattern of all lower extremity muscles when looking at functional leg length differences. In most cases, observations include a lateral displacement of the pelvis away from the short leg, allowing for hip abduction in the apparent short leg, and tightness in the associated muscles. The abductors on the longer leg become weakened. In many cases, some re-education of the gluteal muscles may be required, as well as postural awareness training.

Are the kneecaps forward facing?

An inward facing or excessively outward facing patella may be associated with medial or lateral hip rotation, respectively. Any observations of medial rotation should be correlated with pronation patterns, weakness in the lateral hip rotators, and an increased lumbar lordosis. Long-term medial rotation with locked knees may also be associated with postural bow legs. Excessive medial rotation of the femur may be due to a structural variation known as femoral anteversion. If present, movement should be administered appropriately, to protect the hip joint capsule from excessive rotation beyond its structural capabilities.

Excessive lateral rotation is often a sign of a structural adaptation of the femur during early growth, and is sometimes referred to as femoral retroversion. There is almost always an associated out-toeing with the lateral hip rotation.

There are some instances where the knee caps may still face forward in spite of an apparent lateral rotation, as observed with out-toeing. This is commonly seen in dancers who require the aesthetic appearance of out-toeing, yet are unable to achieve the full degree of hip rotation to perform this. In this case, lateral tibial torsion provides the motion for a restricted hip joint. This may contribute to laxity in the cruciate ligaments, and/or meniscus problems, as the knee attempts to complete the rotation instead of the hip joint. Postural movement that focuses on improving control of hip joint rotation may be beneficial.

Are the kneecaps tracking centrally over the knee joints?

A medially tracking patella may be the result of a muscle imbalance in favour of excessive pull by the vastus medialis; a laterally tracking patella is indicative of excessive pull or hypertrophy of the vastus lateralis. In either case, the extent of tracking can be determined by measurement of the Q-angle, as well as more subject-ively through the observation of apparent hypertrophy or atrophy of the vasti muscles. In these instances there is likely to be associated deviations at the ankle and hip. While specific exercises may target the vasti muscles alone, an integrated approach that focuses on restoring lower limb alignment, foot/ankle control, and hip joint control will improve longer-term postural health.

The thigh and femur

Is the muscle mass and tone of the quadriceps muscle group similar and appropriate?

Excessive hypertrophy and/or tone in the quadriceps may be indicative of altered dominance patterns due to footwear (such as wearing high heels); the repetitive performance of task or sports-specific movements (such as running); faulty exercise habits; or injury or trauma to a functional antagonist or synergist, resulting in inhibition of that muscle.

Figure 3.12 Comparison of tone and muscle mass in the thighs

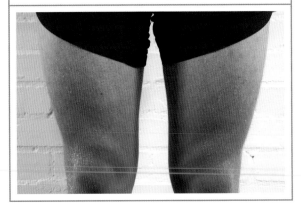

One-sided differences in tone and mass may suggest dysfunction of the knee joint or lumbo-pelvic hip complex, as well as possible imbalances in vestibular function. It's important to understand that unilateral hypertrophy exists in many activities and sports, and should not always be regarded as posturally unsound. However, if such differences are contributing to pain and dysfunction, then intervention may be required. A number of single leg movements can be used to effectively target unilateral muscle imbalances, as well as improving control of balance; the vestibular system may be challenged further using oculo-vestibular exercises.

The pelvis

Are the anterior superior iliac spines (ASIS) level?

Elevated anterior superior iliac spines (ASIS) that is, the anterior extremity of the iliac crest of the pelvis, on one side may suggest weakness of the same side hip abductors, which is often associated with a dominance of the same side quadratus lumborum. The opposite pattern will usually occur on the other side. This can give rise to a scoliotic pattern in the spine and the appearance of a short leg on the opposite side. It may also indicate a rotational dysfunction with the innominate bone of the elevated side tilting posteriorly, and the depressed side tilting anteriorly.

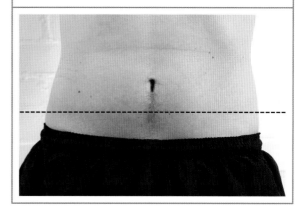

Figure 3.13 Observation of the pelvis

The dotted line shows the top of the pelvis.

With a minor deviation, a heel lift placed under the short leg should level the ASIS and straighten out the spine, if the spine is not rigidly fixed. If the deviation is large, chronic, or structural, then a significant correction may not be achieved with a heel raise, and subsequent referral to a manual/manipulative therapist may be required for postural re-education.

When viewing asymmetries of the pelvis, it is important to consider the compensatory patterns that occur up and down the body. Hip hikes result in a shift of body weight laterally, away from the side of the hike. To compensate, the same side shoulder may drop; if left uncorrected, the neck may assume an opposing movement to preserve the horizontal position of the eyes. The result is a lateral S-shaped curve or scoliosis, which may be functional or structural.

Because hip-hiking and scoliotic patterns are often associated with muscle shortness in the trunk muscles, stretching movements are often used, as well as manual therapeutic techniques such as massage. Where there is muscle tightness, a number of release techniques that restore the muscle to lower levels of tone, followed by mobility exercises are also recommended.

The trunk

Do the ribs appear balanced centrally over the pelvis? Do the ribs appear excessively flared out, or well hidden?

While the overall shape of everyone's ribcage and ribs is different, the sternum should be positioned over the symphysis pubis, with balanced flare of the ribs on both sides.

Figure 3.14 Observation of the ribs

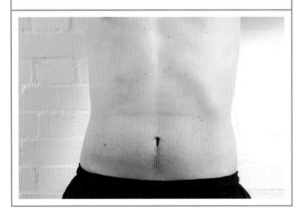

Deviation of the sternum from a central position may suggest a scoliotic pattern; differences in flaring of the ribs may be measured by the sterno-costal (SC) angle. A decrease in the angle may suggest shortness of the upper fibres of the rectus and external obliques, as well as the pectorals and latissimus dorsi. Differences in the SC angle may also be associated with faulty breathing mechanics, and may also occur unilaterally (as in scoliosis).

Mobilisation of the ribs coupled with release techniques for the abdominal and hip musculature may help to restore optimal ribcage alignment over the pelvis, as well as provide a way for improving breathing technique.

Do the ribs exhibit full excursion during breathing?

There should be some expansion of the (lower) abdomen during relaxed breathing, with a small amount of lateral movement in the lower ribs. With a deeper inhalation there should be a definite lateral/superior/anterior movement of the rib cage as the thorax expands.

Is the distance between the lower ribs and the ASIS equal on both sides?

Shortness on one side may be associated with shortness of the quadratus lumborum, latissimus dorsi or obliques; there may also be a scoliosis.

When standing, such a shortness may not exert a large enough pull to elevate the pelvis, but instead may be apparent by an observable depression of the ribs, and sometimes a downward pull of the shoulder; visible elevation of the pelvis may become more apparent when lying supine, as the quadratus muscle in particular assumes a shorter position, due to its reduced effort in pulling the pelvis.

Does the upper portion of the abdominal muscles appear to be balanced with the lower portion, in relation to tone and size? Do the oblique muscles appear to be more prominent or well developed compared to the rectus abdominis?

A common postural deviation is excessive shortening of the fibres of the upper rectus, which may be due to postural adaptations (such as prolonged sitting in a forward or slumped position), or the excessive use of trunk curling exercises. Lack of attention to lower abdominal training may result in their lack of tone and/or atrophy.

Poor tone in the lower abdominal muscles may indicate inhibition by hypertonic lumbar extensors or hip flexors. During supine trunk curl exercises, dominant hamstrings may contribute more to pelvic stability instead of the lower abdominals, thus weakening the lower abdominals over time.

Dominance of the oblique musculature can occur due to training adaptations (such as swimming), poor exercise choice or through the demands of occupation and daily life.

While specific movements exist for isolating the abdominal muscles, better posture will result by improving control of the trunk flexors in relation to one another, as well as integrating their function with that of the back extensors. Once control has been restored and enhanced, further improvements in functional strength will be gained through everyday activities.

Is the 'key-hole' (the distance between arms and trunk) equal on both sides?

Unilateral (one-sided) differences in the size of the 'key-hole' may suggest scoliotic patterns, which may also be signs of shoulder and pelvic dysfunction, leg length discrepancies, and knee and ankle dysfunction.

Figure 3.15 Observation of the 'key-hole'

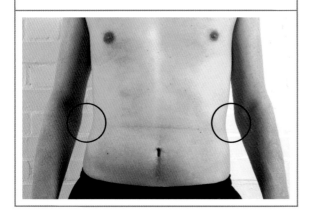

The circles denote the 'key-hole'.

The shoulders and arms

Do the shoulders appear level?

Excessive elevation may suggest hypertonicity of the upper trapezius or levator scapulae; in the case of the levator scapulae, there may also be shortness of the rhomboids. Shortness of the levator scapulae alone may give rise to a distinct elevation of the superior angle of the scapula, causing the inferior angle to move medially.

Figure 3.16 Observation of the shoulders

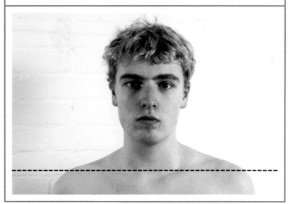

The dotted line shows the level of shoulders – note that the subject has uneven shoulders.

An elevated shoulder is often seen alongside an elevated contra-lateral ASIS (opposing hip hike), as the body attempts to right itself when weight shifting. Because hip hiking is consistent with a lateral curvature of the spine, the opposite shoulder elevation may be part of an overall scoliotic pattern.

Level shoulders are generally a rarity among adult humans, due to handedness, and minor deviations are not usually problematic. Where pain or dysfunction is present, a number of postural strategies exist that integrate shoulder and hip function with the spine, providing improvements in postural control, and pain resolution.

Are the elbow creases forward facing?

Excessive inward facing of the elbow creases suggests an imbalance of the rotator cuff muscles in favour of the medial rotators. This observation may also be consistent with a flexion posture (upper crossed syndrome).

Figure 3.17 Observation of the elbow creases

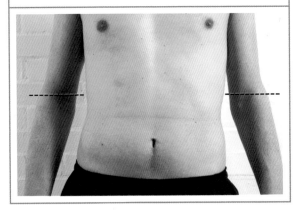

The dotted line denotes the level of the elbow creases.

However, the larger mechanically suited muscles of medial rotation, including the pectoralis major and the latissimus dorsi, should also be considered as contributors. It is also important to observe any pronation/supination occurring at the forearm; often these imbalances may be incorrectly taken for rotator cuff imbalances.

Is the tone and mass of the upper trapezius and deltoid muscles evenly balanced?

Hypertonicity of the upper trapezius will almost always be associated with elevation of the scapula, and may be associated with poor scapula control when seated and standing; unilateral hypertonicity is usually associated with handedness.

Figure 3.18 Tone, size and symmetry of the deltoid and trapezius muscles

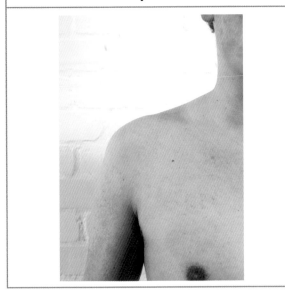

Atrophy of the deltoid may occur as a result of impingement of the brachial plexus, and may also be associated with dysfunction in the lower vertebrae of the cervical spine.

It's important to note that the upper trapezius and deltoid are two important muscles that are part of the force couple responsible for scapulo-humeral rhythm; dysfunction in either will invariably result in poor mechanics of the shoulder complex and/or loss of range of motion. Scapulohumeral rhythm can easily be restored using a combination of release techniques, stretching, and movement awareness. Once restored, functional strength may be increased through the performance of simple pushing and pulling movements.

The head and neck

Is the head evenly balanced on the shoulders?

The following variations may be present when observing head position from the anterior view:

lateral tilt, rotation, lateral glide, or a combination of these variations. It is important to note that a mild degree of cervical distortion can

Figure 3.19 Observation of head position

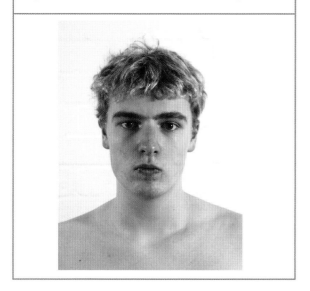

occur naturally due to handedness. Excessive lateral tilt is often associated with stiffness and/or pain in the lower cervical vertebrae, and elevation of the shoulder. The tilt may be more pronounced in the supine position whereas, when standing, the shoulder elevation may be more pronounced.

The presence of rotation may also suggest handedness and/or repetitive habitual use of specific movement patterns or postures, such as holding a phone to the ear. There may also be pain/discomfort associated with the middle part of the neck, with possible trigger points in the levator scapulae, sternocleidomastoid and upper trapezius. The presence of lateral glide may suggest dysfunction of the upper cervical vertebrae, and may be associated with stiffness and pain in this region.

It is important to understand that any cervical distortions may also be the result of, and may

result in, visual-vestibular imbalances, which should be assessed and subsequently addressed during treatment. There are a number of useful vestibular rehabilitation exercises and the practitioner should seek the advice of an appropriately qualified specialist. Release techniques for the neck muscles may also be effective, especially when combined with neck mobility movements, and postural awareness training.

Posterior observations

The feet

Are the Achilles tendons vertically aligned?

A lateral concavity of the tendon will be observed if the foot is pronated, and a medial concavity if the foot is supinated. Extreme deviations may cause knee pain, Achilles tendonitis, and eversion/inversion ankle sprains.

Figure 3.20 Observation of the achilles tendon

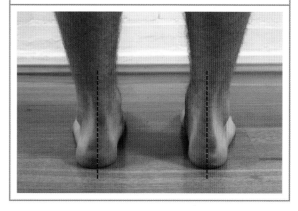

The dotted lines denote the vertical alignment of the Achilles tendons.

The lower leg

Are the calf muscles balanced in size?

Differences in muscle mass in the left and right calf are usually associated with chronic uneven weight bearing that may be the result of unilateral injury, nerve damage, structural variations, or vestibular dysfunction. A small degree may also be associated with unilateral exercise and sport, and should be considered as normal in the absence of pain and musculoskeletal dysfunction.

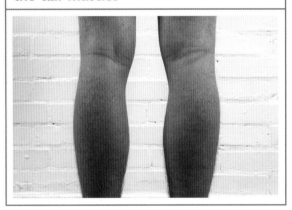

Figure 3.21 Tone, size and symmetry of the calf muscles

Excessive hypertrophy of the calves can be caused by faulty footwear, prolonged exposure to specific footwear (such as high heels), or excessive sporting activity (such as running). It is important to consider the implications of calf hypertrophy and tightness to the health of the surrounding structures, such as the Achilles tendon and ankle; calf tightness is also a probable factor in the aetiology of low back pain and, in particular, its association with excessive lumbar lordosis. Techniques for stretching the calves should also be combined with re-education of the other extensor muscle groups, including the hamstrings and back extensors, for long-term postural health.

The pelvis and hip

Are the iliac crests level?

Refer to the notes on the anterior view.

The spine

Do the cervical, thoracic and lumbar regions of the spine appear vertically aligned?

Lateral curvatures (C-shaped or S-shaped) are associated with fallen arches in the feet, structural or functional leg length discrepancies, pelvic distortions (hip-hike or torsion), muscular imbalances of the paraspinal muscles, and possible vestibular dysfunction. It's worth noting that minor lateral curvatures are common results of handedness.

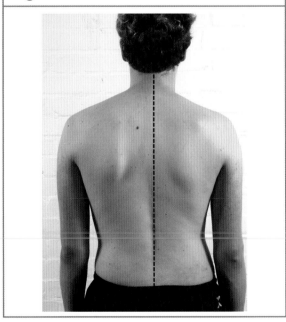

Figure 3.22 Observation of spinal alignment

The dotted line denotes the alignment of the spine.

Lateral curvatures may be structural or functional in origin, though such a diagnosis should only be made by a qualified specialist.

Do any of the spinous processes protrude excessively? Do the spinous processes appear to be rotated?

In the first instance, the presence of a 'naked spine' (protruding spinous processes) may indicate an excessive anterior–posterior curvature in that particular region, and should be regarded as such in relation to the other curvatures of the spine, and whether there is any pain present. Protruding vertebrae may also be present in those people who have very little muscle mass in the paraspinal muscles, possibly due to atrophy or lack of use. In these cases, exercises that help to restore muscle mass may meet rehabilitative and aesthetic objectives.

Where scoliotic patterns exist, closer observations (and palpation) of the spinous processes may reveal rotation or 'kinks', and the vertebrae do not appear to line up vertically. Side bending of the spine is coupled with rotation to the same side or opposite side, depending on whether the spine is in neutral or not. In any case, the rotation/side bending may be associated with facet joint restrictions that may require releasing by an appropriately qualified practitioner. Once released, health of the facet joints can be maintained through postural movements that focus on all aspects of spinal mobility.

Does the size of the paraspinal muscles appear balanced from left to right? Is there excessive hypertrophy/hypertonicity of the thoraco-lumbar musculature?

Observations of asymmetry may highlight important muscle imbalances. It is important to understand that minor asymmetry in muscle mass is allied to handedness, and is even considered to be functional by many experts. A large degree of asymmetry may be associated with muscle imbalances as a result of overuse (or underuse), atrophy due to injury or trauma, as well as spinal, pelvic and foot/ankle distortions. Athletes involved in sports that are predominantly unilateral may have asymmetry.

Excessive hypertrophy of the thoraco-lumbar musculature may suggest abnormalities in the gait cycle or inhibition of the hip extensors (as a compensatory mechanism). It is also an adaptation in power athletes such as power lifters and gymnasts, who rely on the mechanical advantage of these muscles to stabilise and powerfully extend the spine.

The anti-gravity role of the paraspinal muscles is controlled by low level tonic contractions, and any hypertonicity of the paraspinal muscles to the level of palpable tightness may warrant further investigation into dynamic postural habits. While unnecessary tightness can easily be relieved through the application of release techniques, long-term postural change and/or the relief of pain will require higher integration of postural awareness training that focuses on moving with less effort.

The shoulder girdle

Are the scapulae positioned flat against the thorax? Are the medial borders approximately 10cm apart? Are the inferior angles level?

In optimal alignment, the scapulae lie flat against the thorax, with very little or no prominence of the borders. Their position is not significantly affected by excessive muscle hypertrophy or conscious postural correction. The position of the scapulae is approximately between the 2nd and 7th thoracic vertebrae, and the medial borders are about 10cm apart. The inferior (or superior) angles are also level in the horizontal plane.

Figure 3.23 Observation of scapulae alignment

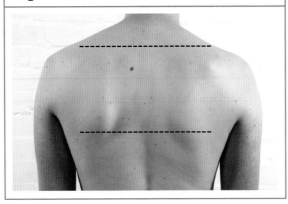

The dotted lines denote alignment of the scapulae, top and bottom.

Differences in the level of the inferior angles may indicate unilateral elevation or upward rotation of the scapula. Excessive scapula adduction may be the result of hypertonic rhomboids, and is often associated with a degree of scapula elevation (upper trapezius and levator scapulae). Scapula elevation often relates to handedness. However, pain patterns and stiffness are commonly associated with prolonged exposure to unilateral elevation, such as using a mouse at a computer, and upper body weight bearing, such as carrying a heavy shoulder bag or brief case. In such cases, the contralateral lower trapezius may hypertrophy. Weakness of the serratus anterior may result in 'winged scapulae', where the medial borders lift away from the thorax.

Because of the many muscle force couples present in the shoulder girdle, re-education and regular practise of scapula and shoulder movements is advised for everyone, regardless of structure and function.

The head and neck

Is the head evenly balanced on the shoulders? Refer to the notes on the anterior view.

Lateral observations

The lateral view offers useful information about joint alignment in relation to the gravity line; for this reason, an overall lateral view should be performed before specific observations are made about individual structures.

Whole body view

From the ground up, is there a vertical alignment of the lateral malleolus, the head of the fibula, the greater trochanter, the head of the humerus, and the auditory meatus, with the gravity line passing through the lumbar and cervical vertebrae?

Deviations away from the gravity line can occur at any level above the lateral malleolus. Moderate shortness in the calves may pull the ankle away from a neutral position in favour of plantar flexion; usually when standing this may result in knee hyperextension (gravity line falls forward of the fibula head) and/or anterior pelvic tilt (lower crossed syndrome pattern); in extreme cases, the person can only stand on their toes – this will usually tension the erector spinae and hamstrings, as well as the toe flexors. In this situation, the greater trochanter will fall in front of the gravity line.

When the greater trochanter lies behind the gravity line, the lumbar spine becomes flattened, and posture becomes ready for weight bearing; the hamstrings are prone to adaptive shortening through prolonged exposure to this posture. This posture is usually associated with an excessive thoracic kyphosis and forward head carriage.

Figure 3.24 Whole body lateral view against plumb line

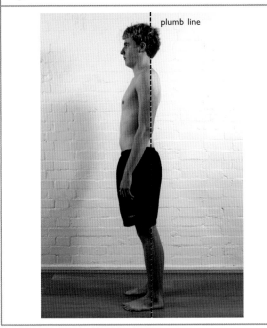

plumb line

Note the subject shows displacement forward of the plumb line.

If the head of the humerus lies forward of the gravity line, there may be medial rotation of the shoulder and/or anterior glide of the humerus. Short or tight muscles may include pectorals, serratus anterior, anterior deltoid, subscapularis, teres major, upper trapezius, and latissimus dorsi. An excessive thoracic kyphosis may also contribute to a forward position of the humeral head in relation to the plumb line.

A forward head carriage is the result if the auditory meatus is forward of the plumb line. Muscles responsible for this positioning include the sternocleidomastoid, cervical extensors, scalenes, pectoralis minor and upper fibres of the rectus abdominis. It's important to note that excessive spinal curvatures in the lumbar and thoracic regions can also produce secondary curvatures in the cervical region. Prolonged seated postures at computers, incorrect sitting posture and short-sightedness can all contribute to forward head carriage.

In an optimal head position, the eye orbits should be forward facing, and not aimed upwards or downwards.

Weight bearing

Is body weight positioned centrally over the arches?

Refer to the notes on the anterior view.

Pelvis

Are the ASIS (anterior superior iliac spines) and PSIS (posterior superior iliac spines) level in the horizontal plane?

The ASIS and PSIS (the posterior extremity of the iliac crest of the pelvis) should be level when the practitioner's fingertips are placed on either side; a common anatomical variation (especially in females) exists where the ASIS is slightly lower then the PSIS.

Figure 3.25 Observation of pelvic alignment

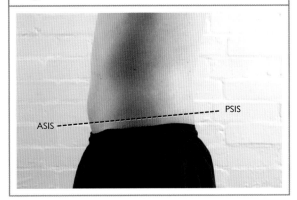

PSIS

ASIS

Note the subject displays a small degree of anterior tilt.

If the ASIS is much lower than the PSIS, the innominate is anteriorly rotated and the sacrum is in anterior tilt, both of which contribute to increased lumbar lordosis. If one innominate is rotated anteriorly and the other posteriorly, the result is pelvic torsion, which may also be associated with a scoliosis.

Spinal curvatures

Does the spine follow a smooth progressive S-shaped curve?

The spine should follow a smooth progressive S-shape curve, without excessive curvatures in the cervical, thoracic and lumbar regions. Any 'kinks' or points of stress may warrant further investigation from the posterior view or when lying prone. The use of a flexible curved ruler may provide a more visual representation of the spinal curvatures.

Figure 3.26 Observation of spinal alignment

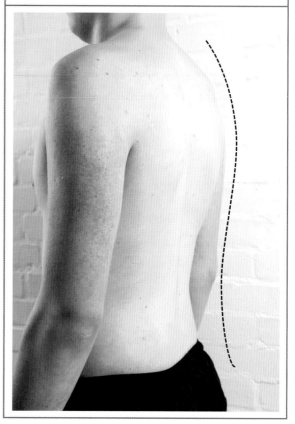

The dotted line denotes the S-curvature of the spine.

Long-term postural health of the spine should involve a number of different movement strategies centred around the incredible movement potential of the spine, as well as how these movements integrate with distal body parts.

Seated and lying assessment

Assessment of seated and lying posture can help to confirm observations made of standing posture, as well as providing areas for further investigation. With many people spending a large part of their day seated, there is also a functional aspect to this assessment. While assessment in a lying position may not necessarily offer information from a functional perspective, it may help to clarify joint alignment in a position where the effects of gravity on the axial skeleton are somewhat reduced; this may help further differentiation of structural and acquired variations in posture.

Seated assessment

Habitual sitting posture often reveals important clues about postural efficiency. This assessment has functional relevance if there is prolonged exposure to these positions. Predominant use of one hand, such as using a mouse, may lead to sustained stress on the cervical spine, shoulder girdle, ribcage and lumbar spine; preferential leaning towards one side of a chair or desk may have similar effects. The presence of pelvic torsion may become more pronounced when sitting compared to standing, as the majority of leg support is taken away, revealing a marked dropping of one side of the pelvis and a scoliotic pattern in the spine. Where prolonged sitting is common, appropriate strategies should be designed to minimise structural stress, by re-educating sitting posture, as well as providing release for tight muscles. Where possible, modifications to chair and desk positions should also be made. Because many different seated positions exist, it is important to correlate seated observations with standing observations. For certain postures, sitting down may be prophylactic, whereas for others, it may exacerbate pain. In the latter case, certain muscles that are already short in the standing position may be subject to stretch whilst seated, contributing further to pain.

The following important questions should be asked when observing seated posture, particularly when the person sits for prolonged periods. Any strategies that they use to remain comfortable in this posture should also be noted.

Observations

How are the feet positioned?

Foot position when sitting can have a dramatic effect on lumbar positioning, and therefore overall spinal positioning. Generally, in an upright seated position, if the feet are placed far away from the chair (sitting with straight legs), the lumbar spine becomes flattened more easily.

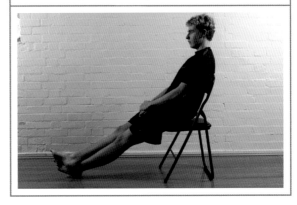

Figure 3.27 Seated position with legs extended, lumbar spine flattened

If the hamstrings are already short, this may pull the lumbar spine into pronounced flexion. If the person is resting against the back of the chair, the lumbar extensors are subject to passive lengthening, or stretch weakness.

Lengthening the hamstrings, as part of a postural re-education programme, may be an important first step, however, instruction in pelvic positioning is vital for long-term correction. Some of this may be helped by instructing the person to place their feet further back towards the chair, and sitting slightly away from the back rest. As their legs move in, the body's centre of gravity comes to rest towards the midline, and the weight of their thighs pulls their knees downwards. This has an effect of passively pulling the pelvis towards anterior tilt, usually with a minimum of effort.

Figure 3.28 Seated position with legs tucked under, and passive pelvic tilt

lumbar spine and slouch, or excessively recruit the lumbar extensors in an effort to sit up straight.

Figure 3.29 Seated position with knees positioned higher than the hip joints

Where a person has an excessive lordosis in sitting, the strategy of decreasing the angle of hip flexion may provide some relief from pain. In this case, placing the feet on a small block may release the tension in the hip flexors, and allow the pelvis to tilt posteriorly.

Is the body leaning over to one side?
Subtle shifts in body weight when sitting can have long lasting effects on posture and can contribute to low and upper back pain. On one hand, the structural effects of a scoliosis may predispose someone towards a position where weight is shifted towards one side of the pelvis. On the other hand, making repetitive side reaching movements when sitting may eventually lead to a similar weight shift, around the body as it re-organises itself, causing a similar postural pattern.

How are the thighs positioned?
When viewed from the side, favourable spine positioning can be achieved easily if the hip and knee joint are positioned in the same horizontal plane, or if the hip is slightly higher than the knee. This position allows potentially more range of motion at the pelvis, particularly if the hamstrings are short. As the angle of hip flexion decreases, so too does the ability to hold the lumbar spine and pelvis close to a neutral position. In this case, the person will either flex their

Figure 3.30 Seated position – leaning over to one side

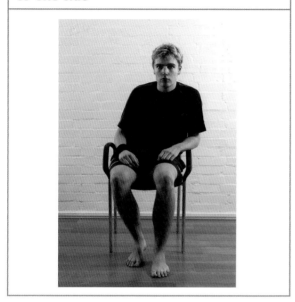

Usually, a change in postural awareness, alongside ergonomic changes, will break this pattern. Regular practise of postural movements that focus on the integration of the opposing hip and shoulder may also be beneficial.

Is the person leaning against the back rest?

Support against a back rest allows the extensor muscles and other paraspinal muscles to relax somewhat, and spine position becomes increasingly reliant on ligament support. As mentioned previously, if the spine is already in a position of flexion, this ligament strain can be pronounced, and over time can lead to deformity and subsequent laxity in these structures as they begin to 'creep'. Where postural strain is due to endurance weakness in the back extensors, use of the back rest is important providing the spine is maintained in an optimal position along with the pelvis. As muscular endurance of the extensors increases through exercise, the ability to sit away from the back rest with postural control will improve.

Lying assessment

Lying assessment involves making observations of the supine and prone (and sometimes side-lying) positions. As most muscles will assume a horizontal position in relation to their fibre alignment, there will also be a significant reduction in their tonus; such a change may help to clarify any potential structural variations that are not clearly seen in standing.

In the supine lying position, the practitioner can further confirm pelvic distortions, as well as getting a feel for the degree of mobility in this region, in the presence of limited muscular constraints. At the hip, the degree of medial or lateral rotation can be observed by the presence of in-toeing or out-toeing. Other observations in the supine position may include rib cage alignment and mobility, patterns of breathing, and head and shoulder positioning.

The prone position is usually reserved for detailed observation of the pelvis, spine and shoulder girdle. In this position, alignment patterns between the pelvic girdle and shoulder girdle can be assessed, and how the spine connects these regions.

In addition to the above observations, both the supine and prone position also allow the practitioner to palpate the muscles easily to detect potential tenderness or pain that may not otherwise be visible or palpable when standing. Areas of increased tone, particularly in the extensor muscles should be queried, and possible reasons as to why these muscles still have a higher level of tone despite the prone body position, should be explored.

The following questions may help the practitioner to confirm some of the findings of their standing assessment, as well as highlight areas for further investigation. This is by no means an exhaustive list of observations, but it should enable the practitioner to begin a series of explorations from which to build postural re-education strategies.

Supine observations

Is the head rotated/displaced?

Rotation of the head to one side may suggest shortness or tightness in the cervical rotators, with the sternocleidomastoid and scalenes as important contributors.

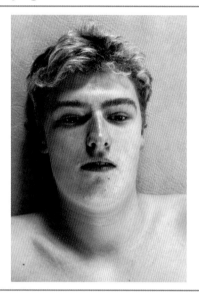

Figure 3.31 Observation of head position when lying

By placing their hands on either side of the person's head, the practitioner may be able to assess any restrictions by gently rotating the head from side to side. In some cases, the head may be rotated or displaced, due to improper positioning when lying supine. When corrected, there may be no restriction of movement.

Are the shoulders elevated? Are the back of the shoulders in contact with the couch?

Elevation of the shoulders in the supine position may suggest shortness in the upper trapezius, levator scapulae or rhomboids. It is important to note that tightness in these muscles may present as shoulder elevation when standing, but when lying, may relax sufficiently to level the shoulders.

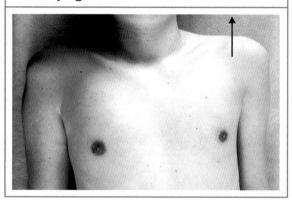

Figure 3.32 Observation of the shoulders when lying

The arrow denotes the elevation of the left shoulder – note that the subject has uneven shoulders, which is consistent with standing observations.

Do the left and right sides of the ribcage appear balanced?

The ribcage may be observed and/or palpated for signs of asymmetry. While some structural asymmetries exist in all individuals, larger deviations may indicate imbalances in oblique strength. In this case, observation of the sternocostal angle will reveal the extent of this.

Figure 3.33 Observation of the ribcage when lying

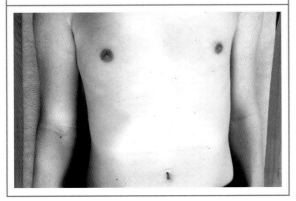

Ribcage asymmetry is also associated with scoliotic patterns (due to spinal rotation), and hip flexor shortness (as the lumbar spine is pulled laterally). Improvements in alignment may be gained by breathing properly through rib mobilisations (by movement sequences or manual manipulation), and general improvements in spinal mobility.

Are the ribs mobile during breathing? Is there equal movement of both sides of the ribcage during breathing? Is there evidence of diaphragmatic breathing? Is there obvious accessory muscle recruitment during relaxed breathing?

While the subject of breathing is vast, a few simple concepts are discussed below, and put into the context of postural health. The assessment of breathing can be performed standing or sitting. The supine position offers an easier route for palpation of the ribcage in the absence of direct gravitational forces, and may provide further information about rib dysfunction.

The following observations can be made during breathing:
- Does the abdomen move anteriorly during inhalation, and to what degree?
- Is there movement in the upper chest during relaxed breathing?
- Is there an obvious lateral excursion of the ribs during inhalation?

Movement of the abdomen during breathing suggests proper use of the diaphragm, while excessive chest movement may suggest an overuse of accessory breathing muscles. Observation of the neck muscles during inhalation may highlight this, as they contract and elevate the upper ribs. Lateral movement of the ribs is evidence of good rib mobility.

If the ribcage is rotated in the supine position, full diaphragmatic breathing may be impeded. Rotation may be assessed by placing the hands around the lower half of the ribs and the fingers along the rib shafts. By viewing the ribcage as a cylinder, the practitioner can gently explore rotation left to right (and also side bending), until a rotational preference is discovered.

Release of the neck, scapula and trunk muscles will often open the doorway to fuller breathing, and may be used effectively in conjunction with rib and spine mobilisations.

Is the pelvis balanced from left to right?

When standing, a hypertonic or short quadratus lumborum may not be able to exert a strong enough upward pull against the pelvis to produce a significant hip-hike in standing; whereas in the supine position, there may be a more marked elevation of the ASIS on the side of shortness.

Figure 3.34 Observation of the pelvis when lying

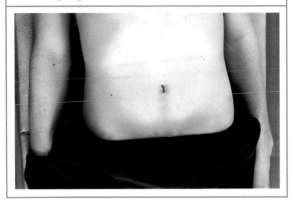

Figure 3.35 Observation of the hip rotation when lying

Any side bending of the lumbar spine will also be consistent with rotation, and this may be evident as one ASIS being higher than the other in the vertical plane. Pelvic torsion may also be present, as observed by an anterior tilt of one ASIS relative to the other. If present, further investigation may be warranted by a suitably qualified practitioner.

Is there evidence of hip rotation?

The extent of hip rotation is often more evident when supine when the feet are no longer fixed. This is easily seen by the degree of out-toeing. Excessive out-toeing may suggest shortness in the lateral hip rotators, and elongation of the medial rotators. To confirm this, observations of thigh orientation should also be made, and in particular, positioning of the patellae. This may provide stronger evidence of hip rotation.

Note that the subject shows lateral rotation of the right hip, as suggested by the out-toeing of the right foot and rotation outwards of the right thigh.

Any hip rotation should be viewed within the context of lumbo-pelvic rotation. For example, if a lumbo-pelvic rotation exists to the right when supine, the right hip will appear to be positioned in lateral rotation relative to the left.

Is there a structural leg length discrepancy?

Objective values for leg length can be obtained by measuring the distance from the ASIS to the medial malleolus. If present, the use of foot raises may be required to alleviate pain and correct lumbo-pelvic mechanics.

Prone observations

Are the scapulae balanced and well aligned?

Although observation of the scapulae when standing will provide adequate information about muscle tightness and control of movement, the prone position will confirm the presence of muscle shortness. It is important to understand that when lying prone, the scapulae will assume a slightly elevated, abducted and winged position under the weight of the shoulder girdle.

Is there evidence of increased tone in the back extensor muscles?

In a relaxed prone position, there should be minimal tone in the back extensors. The presence of muscle shortness in the thoracic spine may be observed as a flattening of the kyphosis; shortness in the lumbar region may be observed as an excessive lumbar lordosis.

Spinal rotation may also give rise to the appear-

Figure 3.36 Observation of the back extensors when lying

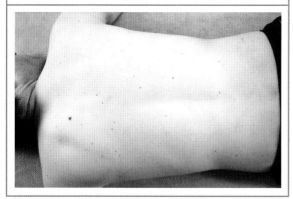

ance of asymmetrical muscle tone. With this in mind, any suspected rotation should be dealt with first, before directly exercising the muscles which are involved.

What is the general mobility of the spine?

The prone position offers a useful position from which to assess spinal mobility, due to the generally lower levels of muscle tone. If the person is relatively pain-free, the practitioner can place both their hands on the sacrum and gently rock from side to side. This type of rocking will assume a rhythmic pattern after a few seconds, and if the individual is relaxed

with little or no restrictions, a corresponding rhythm will also be observed in the lower extremities, as well as the upper back and shoulders.

Figure 3.37 Assessing general mobility of the spine

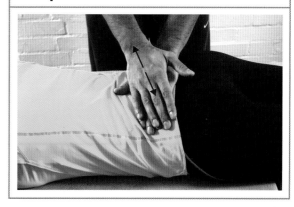

The arrows denote the oscillatory movement of the hands left to right.

As well as being a valuable movement to increase passive mobility of the entire spine, it also serves as a means of assessing how much motion potential the spine has in the first place. Any areas of the spine which exhibit poor mobility or apparent restrictions should be correlated to other observations and/or personal history, and treated accordingly using movement-based strategies that focus on spinal mobility in all planes of motion.

ASSESSMENT OF FUNCTION

4

Introduction

The close relationship between posture and the locomotor system strongly suggests that posture isn't just a static concept. While the observation of standing posture is a useful starting point for understanding how we interact with gravity, we mustn't forget that it also allows us to speculate how such interactions will influence function through movement. Through all of this, we should remember that there is no such thing as bad posture – posture should be more suitably regarded in terms of how efficiently we organise our bodies statically and dynamically.

Having already assessed how the body organises itself in relation to gravity when standing, sitting and lying, we are now ready to explore how the body organises itself in preparation for and during movement – a concept that is also known as functional efficiency. To do this, there will be a strong focus on how different muscles and joint systems interconnect; when they connect in an optimal way, movement becomes efficient.

The question of what constitutes efficiency is an important one. Subjectively assessing functional efficiency does not require much more than a good degree of sensory acuity and knowledge of functional anatomy. It requires a keen eye to be able to spot habitual behaviour accurately and translate this into observable movement patterns for assessment. While many different movement tests exist, only a few will yield significant information about posture, and further confirm the findings of a static assessment. While rigorous testing is rarely required, four important questions should be asked during observations of movement:

1. Is the movement being performed with minimal effort?
2. Is the movement well-coordinated?
3. How well does the person return to a position of balance and stability?
4. Can the person stop the movement at any time, and reverse or continue?

When considered alongside the results of structural assessments, the answers to these questions will provide insight of how the body organises itself in preparation for and during movement.

The aim of this chapter is to explore efficiency of movement across a number of fundamental motor patterns that are essential to overall postural health. While human movement is much more diverse than this, it will provide the practitioner with useful information from which to build effective postural re-education programmes, as well as presenting other areas for further testing.

Observing movement

Learning to observe movement is a skill that comes with time and practice. Although the goal is to assess how efficiently a person moves, it is also important for the practitioner to use this time to increase movement awareness for the foundations of re-education. Increasing awareness of how a person currently moves is a key factor in setting a reference point from which to improve postural control. By highlighting areas of efficiency, the practitioner can reinforce existing motor strategies; by focusing on areas of inefficiency, the practitioner can make the individual aware of areas for improvement. This process of feedback is essential for learning. The context for learning can be further enhanced by creating the correct motivation for movement. Often, movement is assessed in a manner that is too clinical and rigid, where certain rules have to be adhered to otherwise the movement is deemed dysfunctional or faulty. Because there are many different ways to perform any given movement, the objective is to observe the process of movement, rather than the goal.

It's important for the practitioner to understand that whatever they choose to assess, and how it will be observed, will only be useful when correlated with the results of previous assessments and historical information. All these results will give the practitioner a wealth of information about how the body is being used, and where improvements in efficiency can be made through a postural re-education programme.

Important questions

Assessment of function is a simple and flexible process. Much of this flexibility occurs through observations of what the practitioner considers relevant and what specific information is required to build a practical postural re-education programme. Initial assessment is important as it provides an essential yet functional reference point from which change can be observed. However, follow up assessment is paramount for continual feedback and programme progression. Two important questions should be asked at this point:
- What to assess?
- How to assess it?

What to assess?

Much confusion surrounds the question of what to test. Should we examine the function of indi-

vidual muscles and joints? Or should we assess complex multi-articular movements? Which movement patterns are the most functional? Should all of them be tested?

For the purpose of simple assessment, some experts have attempted to categorise movement into easily observable sub-movements (Chek 2000; McGill 2004; Cook, 2003), each of which can be assessed in their own right. These include:
- Squatting
- Lunging/stepping
- Pushing
- Pulling
- Trunk rotation

Rationale for testing

The rationale behind testing lies in the observation that all human movement is made up of these sub-movements, mostly in an upright position; the assumption is that if a given movement pattern is inefficient, then its associated sub-movements can be re-educated and absorbed into the original movement pattern, which then becomes more efficient. This type of approach has proved to be successful in many cases, and has gained immense popularity among trainers and therapists trying to change the movement potential in their clients.

In a clinical setting, these movement patterns can be observed easily by way of a movement screen, a process that involves the observation of key functional movement patterns. Although the movements may not completely mimic the functional demands imposed on an individual, the pattern itself may be familiar enough to use similar muscle recruitment patterns. At this early stage, initial observations of ability/skill, as well as joint alignment, can be made. Once basic performance has been qualified, specific functional variations of the above movements can be applied to further test the efficiency of the musculoskeletal system. These may include manipulating repetitions, load, balance and tempo of movement in accordance with how the individual performs these tasks during daily living. This is an important step that is often overlooked once assessment of joint alignment is made. Where pain prevents an individual from performing a movement screen the practitioner may be able to adapt the movement to a suitable level while still preserving the mechanics of the original movement.

Limitations of testing

This type of approach to postural health has certain limitations in relation to efficiency. Firstly, for standardisation of assessment to exist, specific norms and ideals are usually created. Any deviation of the movement away from such norms is considered dysfunctional. It is interesting to note that many of these ideal standards have a strong basis in the maintenance of joint alignment during movement, especially spinal alignment. Where joint alignment deviates from the standard, the individual is often instructed to engage or activate specific muscles to maintain good alignment. While this certainly isn't wrong, it may still require excessive amounts of effort that will contribute to inefficiency and premature fatigue. A far more efficient solution would be to re-educate joint alignment, so that the body has to rely less on excessive muscle recruitment. If the body learns to use minimal effort for any given task, it will consistently learn to apply this.

Secondly, this type of approach is often biased towards optimal muscle recruitment, instead of optimal force transmission. When the body learns to organise its parts in relation to one another as well as in relation to the external environment (especially gravity), efficient force transmission through the chain can occur with little effort. This intimate relationship between proprioception and exteroception is often missing during assessment, where emphasis is placed on the goal rather than the process of how the body is being used.

Lastly, many of the above movement patterns do not bear a strong relation to many everyday functional movements such as walking, sitting, reaching, or rolling. For example, good execution of an overhead press movement does not guarantee functional efficiency during overhead reaching; optimal strength in the trunk rotators has little to do with the ability to roll; and good squat technique may not improve the ability to sit down or stand up from a chair. Therefore, in addition to the above basic movements, the practitioner may wish to explore other functional patterns, especially those which the individual finds challenging.

How to assess?

Observing *how* well the body organises itself in preparation for and during movement involves more than a simple observation of joint alignment, although this will provide a good starting point. Successful assessment will also include observations of muscle recruitment, link sequencing and perceived effort, all of which can be subjectively observed and assessed easily, without the need for special equipment. Objective data collection is rarely required, with the emphasis being placed on visual, auditory and kinaesthetic feedback instead. Some key considerations of what to look for during assessment are discussed below.

Movement involves minimal effort

The amount of effort involved during movement depends on a number of factors, and while it is a very subjective observation, it is also one that can give the practitioner a lot of information.

Many people relate effort to fatigability, a concept that is associated with suboptimal cardio-vascular endurance and anaerobic capacity, poor nutrition or lack of sleep. While this is true, particularly during prolonged movement, the idea that suboptimal efficiency can increase physical effort should not be overlooked. When observing a movement pattern, a lack of strength in the first instance may result in increased effort. Similarly, as the movement continues, efforts may increase as muscular endurance fails. In both instances, if the movement is to continue, compensatory movement may take place. While this may serve to maintain the movement, it invariably results in one that is less efficient. Both strength and endurance (as contributors to efficiency) can be assessed functionally by accurate manipulation of resistance and repetitions during any given movement.

If joint alignment is less than optimal during movement (in relation to the gravity line), then certain muscles will be required to work harder in an attempt to maintain stability and balance. While this may be acceptable in the short term, it can lead to muscular pain and joint dysfunction over time. A simple example of this is during forward bending (trunk flexion) with straight legs. This movement projects the body's centre of gravity forward and, to remain balanced, there is excessive recruitment of the lumbar extensors. Prolonged exposure to this type of movement may lead to strain patterns in the low back and disc pathology. Apart from the obvious observations of instability during movement, the practitioner may also ask the subject to perform the movement against a plumb line.

Movement is well coordinated

The observation of well-coordinated movement is a good indicator of neuromuscular efficiency. This often manifests as a movement that appears to involve integrated use of body parts, in a manner that is smooth and fluid. While 'jerky' movement may be a sign of physical fatigue, in the presence of sufficient energy resources they may suggest 'nervous' fatigue. The integrated use of body parts to effectively transmit force through the kinetic chain is a major contributor to overall efficiency. Although a specific movement pattern can be performed using an almost infinite number of joint configurations and muscle recruitment patterns, only a few will result in the best economy of effort. Usually this will occur when the muscles involved exhibit optimal length–tension relationships and balanced force couple action. The presence of altered dominance patterns due to weakness or inhibition of agonist muscles will affect overall efficiency by increasing the muscular effort of fewer muscles. While the overall energy expenditure may still be the same, these muscles are prone to overuse and premature fatigue, leading to inefficiency (and possible injury) over time. Dominance patterns will also create suboptimal firing patterns in muscles, many of which can be observed and palpated during assessment. If muscles responsible for movement around a joint engage before those responsible for stabilising that joint, then joint instability may be present throughout the movement, which can reduce efficiency of movement, either through premature fatigue, or compensatory movement.

People who have good body awareness and proprioception often have better control of movement, an adaptation that comes with experience and good use of the body. In the early stages of learning a new movement pattern, muscle recruitment often follows a pattern of reciprocal activation, or co-contraction. Although this may be essential for joint stability in some instances, excessive use often leads to increased effort and lack of coordination. An example of this is commonly observed when an individual stands on a balance board for the first time; most of the major muscles of the body contract to the perception of an emergency situation, and a large proportion of conscious attention is directed towards remaining upright. While this is not an incorrect configuration, it is certainly an inefficient one that expends a lot of energy. As the skill of balancing improves over time, the person will learn the mechanisms for reciprocal inhibition, a process by which functional antagonists are inhibited, as the prime mover muscles contract. This adaptation allows for a large economy of effort.

Easy return to a position of balance and stability

A sign of functional efficiency often overlooked is the ability to return the body to a point of equilibrium, or postural homeostasis, quickly after movement. This is an essential function, as it allows the body to prepare for further movement using the least amount of effort. While this mechanism of control is often taken for granted, it can easily be observed during simple movements, such as squatting or lifting. Repeated signs of instability at the end of a movement may suggest poor coordination and control. The world of sport contains many examples of how this mechanism works well, including the tennis player who quickly resumes an open stance following a cross court return or the gymnast who can rapidly return to a static standing position following a series of somersaults. Although these are highly refined skill sets specific to sport, there is no reason that similar levels of control cannot be attained in postural re-education programmes.

Reversibility

If the maintenance of equilibrium is considered to be an important indicator of efficiency, then the return to equilibrium at any suitable point during movement is equally so. Such a concept is important because of its roots in self-preservation. When predatory animals are searching for food, they often move in a way that allows them to stop dead at any moment, while remaining perfectly balanced, as is beautifully exemplified by all types of cats. It is a skill that is effectively used by human trackers today. While the pressures of hunting and gathering are not applicable to us nowadays, the reversibility principle undoubtedly played an important part during our childhood development. This is clearly observable in babies who will crawl uninhibited, then suddenly pause, before continuing the movement. It has been suggested that such behaviour teaches us fine motor control and enhances proprioceptive input in response to gravity, as well as in response to the forces experienced as the feet contact the ground (ground reaction forces). This principle can be observed as an individual sits down on a chair, and then stands up. Optimal organisation of the body in relation to

gravity will allow the movement to be stopped at any point, and subsequently continued.

Movement assessments

The following assessments are more than simple observations of movement. They look at elemental motor patterns with the sole focus on how the movement is performed. There is no distinction between which muscles are used or how many joint actions take place; there is also no distinction between how many individual sub-movements are involved. Where possible, each movement will generally have a particular motivation behind it, to enable the practitioner to observe the movement as an uninterrupted action.

It is important to understand that these movements are referred to as 'elemental' because they are the foundations for human movement; all of them relate to landmark movement patterns we once learned in the first two years of life. These movements helped us to learn how to turn, twist, climb, reach, crawl, sit, stand and walk – observations of these may highlight fundamental faults that a more clinical evaluation may miss.

Where someone is unable to perform a movement or shows a limitation in one aspect of the movement, or demonstrates a difference in function on one side of the body compared to the other, a significant piece of information about that individual will be uncovered. More often than not, the reasons behind this dysfunction are not always due to strength or flexibility deficits, but due to lack of control. Therefore, efforts to improve the movement should focus on improving control in the first instance, rather than strength of individual muscles or body parts.

Objectives of assessment

Many of the following movements require good control of posture (especially as more and more joints and muscles become involved), and this control is dependent on optimal flexibility, mobility, balance, strength, and coordination. Therefore, assessment of these movements will provide the practitioner with useful information relating to these factors. The ultimate objective of the assessments is not to work on the individual deficits that are observed during the movement, but to find ways to improve overall control and efficiency. One such way to improve control of posture is to provide the body with the right type of, and sufficient amounts of, motivation to perform a movement. Once control of posture is realised, factors such as functional stability and strength will inevitably follow through daily activities.

No movement assessment is set in stone, nor should it be considered the gold standard. Human movement is infinitely varied, and the following assessments represent just a few ways in which the practitioner can create movement programmes that re-educate postural control. While it's not important to perform all the assessments, it is important to select the ones that are most appropriate to obtaining the information required at any given time. This selection will be supported by the individual's history as well as the results of static assessment.

Observations

The tests begin with observations of simple movements involving minimal joint action, and progress towards more complex integrated patterns that involve sequential multiple movements, which further challenge stability and mobility. Almost all of these movements are highly functional, not in the sense that they necessarily look like daily movements we perform, but rather in the way that the different body parts configure themselves in relation to one another during the movement. This represents an important shift away from the achievement of a movement, towards a mode of assessment that focuses on how a movement is performed – in other words, efficiency of movement.

Almost all these movements involve body-weight only, with no additional loads, and can be performed by many individuals. They include the following patterns:

- Spinal-pelvic motion
- Scapulohumeral rhythm
- Pushing and pulling
- Squatting
- Lunging and stepping
- Walking gait

While the performance of these patterns is possible for most individuals, their achievement alone is not necessarily indicative of functional efficiency. The practitioner should look at the ease with which the movement is performed as well as compensatory or unnecessary muscle action. If the person is in pain or discomfort, then the tests should be avoided, until an appropriately qualified professional says it is safe to continue. The practitioner should understand that it is not important to conduct all these tests, just the ones that will provide useful information at the time of testing. Observations of posture during movement can be made at any point during a postural re-education programme, and further tests may be performed as the need arises.

Finally, many practitioners may be puzzled by the lack of specificity of some of these patterns to sports-specific movements. What should not be forgotten is that the foundation of all human movement (including sport) is based on the ability to function efficiently – that is, without limitation or restriction. The characteristics of postural control can be easily observed in any movement, and once they have been educated or re-educated, they can be applied to many sports-specific movements with the maximum of ease.

Health considerations

All the movement assessments require adequate levels of mobility and stability, with a minimal need for strength. Therefore, they will be suitable for almost everyone. If a person is in pain or discomfort, then the tests should be avoided, until a qualified medical professional says it is safe to do them.

Test 1: Spinal-pelvic motion

Purpose

To observe and assess the relationship between the spine and pelvis when performing the trunk movements of flexion, extension, lateral flexion and rotation.

Action

The subject stands with their arms by their side, and performs the following movements against a plumb line. The practitioner observes spinal and pelvic motion.

- Flexion – the subject slowly bends forwards, as if touching their toes, moving only as far as is comfortable.
- Extension – the subject slowly bends backwards, moving as far as comfortably possible.
- Lateral flexion – the subject slowly bends to one side, allowing their arm to slide down the lateral thigh, and repeats on the other side.
- Trunk rotation – the subject places both arms across their chest and gently rotates their trunk left and right.

Observation

During flexion, efficient use of the body will result in a distinct rhythm between the lumbar spine and pelvis. As the subject bends forward, the first motion will occur at the pelvis, as it tilts anteriorly to prevent the lumbar spine from flexing. The cervical spine may also flex to turn the head and eyes towards the floor. As the hip joint reaches the limits of hip flexion (primarily determined by the existing hamstring length), the spine will begin to flex progressively, with most of the movement occurring in the cervical and thoracic segments; the lumbar spine should remain flat or close to neutral.

If a particular movement requires the person to bend further, the knees will usually bend, or

Figure 4.1 Observation of the spine and pelvis during flexion

the spine may flex further; either of these strategies will be useful. As the subject returns to an upright position, the pelvis will once again initiate the movement, by posterior tilting to engage the gluteal muscles; there may also be a pulling back of the upper arms as the latissimus muscles provide stability to the middle/upper back. The spine progressively restacks once the posterior tilt has been initiated, and the body returns once again to a neutral standing posture.

It's important to note that at the bottom of the movement, there should be a progressive curve in the spine, with most of the curvature occurring at the thoracic and cervical segments. When viewed against the plumb line laterally, there will be a distinct posterior displacement of the hip joints, as the trunk bends forwards, in an effort to maintain balance.

Limited range of motion may suggest muscle shortness but, for adequate function, compensatory motion is likely to become established over time. For example, short hamstrings may be compensated for by an increase in the length of the thoracic extensors (or lumbar extensors) to preserve range of motion in bending. This can contribute to postural dysfunction and/or back pain, particularly if overused.

Lack of anterior pelvic tilt may also be indicative of short hamstrings or simply poor control. On the other hand, tight lumbar extensors and elongated hamstrings may inhibit posterior pelvic tilting, resulting in excessive use of the low back.

During extension, a number of similar observations will occur that indicate efficient use of the body.

maintain optimal stability and range of motion for the spine.

Shortness in the abdominal muscles and/or weakness in the gluteal muscles may result in shorter range of motion in extension, although increased extension in the upper thoracic spine may compensate for this; it may become a source of back pain if overused as a strategy for extension, potentially resulting in facet restrictions and muscle shortness.

Lateral flexion is best viewed posteriorly against a plumb line.

Fig 4.2 Observation of the spine and pelvis during extension	**Fig 4.3 Observation of the spine and pelvis during lateral flexion**
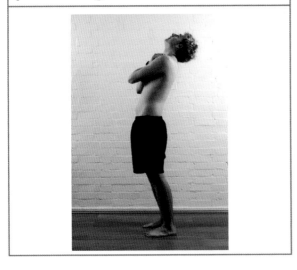	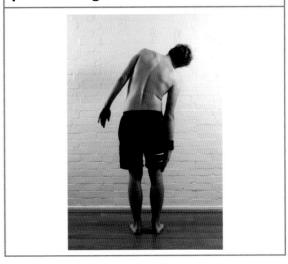

There will be a progressive curvature in the spine (albeit to a small degree), with most of the motion occurring in the cervical and thoracic segments. There may also be an initiation of the motion with the cervical spine, as it rotates the head backwards; however, many people will reflexively maintain a horizontal eye position to maintain equilibrium and balance. A forward displacement of the hips will effectively counterbalance the torso, and a posterior tilting of the pelvis (via the gluteal muscles) will

As previously discussed, optimal use of the body will allow the pelvis and spine to interact rhythmically. As the subject side bends, there will be an opposing movement of the contralateral hip to maintain balance – this will be seen clearly against the plumb line. Movement should be compared left to right to determine any differences. A view of the spine will highlight a lumbar region that exhibits very little lateral flexion, with most of the movement occurring in the thoracic and cervical region. There will also be a progressive spinal curvature.

If the muscles on one side of the spine are short/tight, there may be limited range of lateral flexion to the opposite side. However, functionally, there is often compensation by an increased lateral displacement of the ipsilateral hip, as seen against the plumb line. There may also be increased flexion mobility in the upper thoracic spine on the opposite side, so that overall range of motion is preserved. These patterns may be part of an existing scoliosis.

During standing rotation, the motion of the spine and pelvis occur in the same direction.

As the trunk rotates to the left, for example, so too will the pelvis. Efficient use of the body will allow the pelvis and lumbar spine to move effectively as one unit. Functionally, further increases in rotation can be brought about by allowing the opposite heel to lift, thus creating a pivot on the ball of the foot. This is an efficient mechanism that is used extensively in sports such as golf, tennis and boxing.

Fig 4.4 Observation of the spine and pelvis during rotation

Test 2: Scapulohumeral rhythm

Purpose

To assess the integrated relationship of the shoulder girdle and shoulder joint (scapulo-humeral rhythm).

Action

The subject stands and performs shoulder abduction with the arm straight. The practitioner observes the movement of the scapula and humerus (visually and by palpation) in relation to timing, range of motion, and muscle recruitment.

Observation

It is important to understand that for every 3° of shoulder movement during abduction, the glenohumeral joint contributes approximately 2° of movement and the scapula contributes 1° of movement. This timing is considered to be the norm for shoulder movement. Scapulo-humeral rhythm is seen in all movements of the shoulder, especially during the later stages of flexion and abduction. It is important to note that if the scapula and humerus are not correctly aligned at the start of movement then compensation will occur during movement giving rise to potential joint and muscle stress; with this in mind, the results of static shoulder alignment should be considered before observing movement in this way. Some important observations and deviations are briefly discussed below.

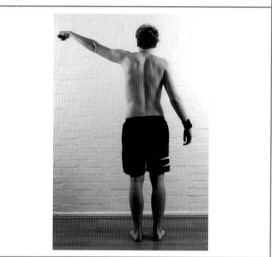

Fig 4.5 Observation of scapulohumeral rhythm

Upward rotation

During full flexion, the scapula will stop moving when the shoulder is flexed to 140°, with the remaining movement occurring at the gleno-humeral joint. At 180° of flexion, the inferior angle should be close to the midline of the thorax in the frontal plane, and the medial border should be upwardly rotated to approximately 60°. Excessive kyphosis and/or a short pectoralis minor can inhibit this end range movement.

Movement of the inferior angle beyond the midline indicates excessive scapula abduction, suggesting rhomboid weakness.

Scapula winging

The scapula should not wing during the movement, or during the return. Winging indicates weakness of the serratus anterior muscle.

Scapula elevation

There should be some elevation (shrugging) of the shoulder during the movement, especially during the later stages (greater than 90°). If there is excessive scapula depression at rest, then elevation of the scapula during these movements must be restored, usually by strengthening the upper trapezius.

Spine

There should be minimal movement of the spine during a full range of movement. A thoracic kyphosis will cause an anterior tilt of the scapula, thereby limiting full flexion capability. This is a movement impairment commonly seen in those who have to sit for long periods of time.

Scapula adduction

During the movement, adduction of the scapula is a sign of rhomboid dominance and/or poor control of shoulder rotation. If there is an accompanied posterior movement of the humeral head, this may indicate dominance of the posterior deltoid over the teres minor and infraspinatus.

Test 3: Pushing and pulling

Purpose

To explore scapulohumeral rhythm further during a standing push and pull movement.

Action

A long piece of exercise tubing or band (low to moderate resistance) is wrapped around an immovable object at chest height, so that the two ends are free. Alternatively, a twin adjustable cable machine may be used.

Pushing: the subject stands with their feet apart, holding both ends of the tubing. Without further instruction, they perform several pushes, so that both arms are straightened in front of their body, returning each time to the start position.

Fig 4.6 Observation of pushing

Pulling: the subject stands facing the tubing, and holds the ends of the tubing in each hand. Without further instruction, they perform several pulls of the tubing, as if bringing both hands towards the body, returning each time to the start position.

Fig 4.7 Observation of pulling

In both cases, the practitioner observes overall alignment, balance and stability, as well as any strategic changes in body positioning.

Observation

During the standing push, the subject will often lean forwards away from the gravity line; while this is important for the completion of the movement, it may also indicate poor control of the abdominal wall in stabilising the trunk, as well as weakness in the pushing muscles (namely the pectoralis major, anterior deltoid and triceps).

Fig 4.8 Leaning forwards during pushing	Fig 4.9 Adoption of a split stance during pushing

Functionally, the strategy of adopting a split stance during the push will enable a more upright body alignment, with subsequent reduced activation of the abdominal wall.

In terms of efficiency, this is a better strategy and one that is almost always adopted in daily life when trying to push a heavy object, for example. If the subject does not adopt a split stance, it may be because the tension in the tubing is not great enough to cause the subject to change stance. Asking the subject to step further forwards may make this happen, as the tension in the tubing increases. The use of a wide stance during pushing may only increase abdominal activation (and vertebral disc compressive forces), at the expense of good alignment and postural control; however, it may still be relevant in certain situations. Many individuals who have strong abdominal walls will exhibit good pushing capability; however, in the interests of efficiency, they should also be shown how to adopt a split stance to perform the same task.

The standing pull displays the opposite pattern of movement, where the major pulling muscles contract (posterior deltoid, rhomboids, trapezius, latissimus dorsi, biceps). Weakness or poor control of these muscles can result in excessive pull of the back extensors (over-arching of the spine), as well as fixation of the scapulae in adduction.

Fig 4.10 Excessive spine extension during pulling

In terms of stability, recruitment of the lumbar extensors is required to counterbalance the pull of the body forwards and, if they are weak, there may be rounding of the back during the movement. People are often taught to engage the muscles of the abdominal wall during a pull; however, this not only inhibits breathing it may also potentially inhibit the back extensors. Abdominal wall activation may only be necessary during heavy lifting.

The postural strategy of adopting a split stance during pushing and pulling not only improves alignment and overall stability in the sagittal plane; it also allows for better lumbo-pelvic-hip mobility, which makes the movement easier through further integration of body parts. During pushing, the individual can now lunge into the front leg, which allows the body's centre of gravity to move forward, thus improving efficiency of movement. The opposite occurs during a pull, where the centre of gravity moves posteriorly onto the back leg, as it bends. In this way, the overall effort of pushing and pulling is more evenly distributed through many joints and muscles, rather than just a few.

Pushing may also be explored from a closed chain perspective, such as pushing up from the floor or even from a chair. The floor push (and variations) form the basis for a number of daily and sporting movements, and is also an important way of getting up off the floor or out of bed. Re-education of efficient pushing is important in postural programmes.

Test 4: Squatting

Purpose

To assess mobility and control at the ankle, knee, hip and pelvis, during a symmetrical squat movement. Squatting is performed in many different ways, all of which can be efficient for the task in hand. For the purpose of simplicity, this test will observe squatting in the context of two important daily movements – sitting down and lifting.

Action

The subject begins in the standing position. To observe how the squat is performed when sitting down and standing up, place a chair behind them and tell them to sit down and then stand up. This may be done several times, as the practitioner observes joint action at the ankle, knee, hip and pelvis/low back, and overall alignment and stability.

Fig 4.11 Observation of squatting when sitting

To observe the squat pattern during lifting, place two lightly loaded shopping bags (or dumbbells) on the floor on each side of the subject. They then pick these up and place them back down again. Observations are made of alignment, joint action and overall stability.

Fig 4.12 Observation of squatting when lifting

Observation

When asked to sit down on a chair, the vast majority of people will simply position their feet somewhere in front of the chair and fall backwards into the seat, which is usually followed by a process of re-positioning to get comfortable. When asked to stand up, the process is not always the reverse of sitting down: the person leans forwards and stands up. If the chair has arm rests, they may push down on these for assistance. While many people can replicate this movement thousands of times without any problem, from an efficiency perspective, this may not be optimal. The fall backwards generally occurs because we know the chair will 'catch' us; it may also occur because we lack control and strength in the muscles that are responsible for squatting.

Some individuals exhibit the opposite, whereby they use unnecessarily excessive muscular effort to sit and stand: one example is where there is a backward thrust of the hips and pelvis, which is counter-balanced by a forward movement of the trunk as the hip flexes.

Fig 4.13 Unnecessary muscle effort in sitting down

This action requires a large contraction of the lumbar extensors; if the extensors are weak, the subject is likely to use the 'falling' strategy, or may experience lumbar strain when sitting down. A similar situation occurs if the hamstrings are short, as they pull the lumbar spine into flexion during sitting. Upon standing, the reverse pattern often occurs, where the lumbar extensors overwork to create anterior pelvic tilt in an effort to provide sufficient drive to stand up. Often this is assisted by the use of arm rests (where present), or by pushing down on the thighs.

While these patterns are satisfactory methods of sitting and standing, the question remains as to how efficient they are. For people with back pain, attention to this process is important, especially if there is a lot of repetition in this movement. As with any movement, there should be no restrictions or limitations; in terms of posture, this translates into minimal joint stress during sitting and standing. For this to occur, due consideration must be given to optimal joint alignment and the position of the body's centre of gravity over the feet – the closer the gravity line is to the feet, the greater the stability; the more vertically aligned our joints are, the less the muscles have to work to maintain joint alignment. If the feet are positioned too far away from the chair, the pelvis must move a greater distance posteriorly to sit down, resulting in potentially excessive muscle effort.

Squatting to pick up a load is different to sitting/standing in two important ways: firstly, there is no rest period at the bottom of the movement (compared to sitting on a chair), and secondly, the body is required to ascend under load, which increases the need for stability. Both these factors rely heavily on mobility, strength and optimal alignment of the lower extremity. For example, shortness in the calves will limit ankle flexion (dorsiflexion) during the descent; this limits knee flexion, and the hip may excessively flex so the upper body can reach the floor.

Fig 4.14 Bending with calf shortness

Fig 4.15 Bending with hamstring shortness

As well as putting the calves under strain, this can also result in additional stress in the lumbar extensors. Similarly, shortness in the hamstrings can restrict hip flexion, which is important to maintain a neutral spine alignment. This may invariably result in a rounding of the lumbar spine as the subject squats to pick up the load.

Another common pattern that is seen during lifting postures is where the legs are kept relatively straight as the spine flexes to pick up the load. Although good mobility may exist in the ankle, knee and hip, weakness or lack of endurance in the knee extensors and hip extensors may result in overuse and strain of the back extensors, and increased risk of disc bulging.

Finally, an overall observation should be made of balance, particularly during the transition where the load is lifted. The level of effort is dependent on the load; however, for the purpose of testing, the load should not be heavy enough to present more than a moderate challenge to the subject. Where any loss of balance occurs, consideration should be given to potential joint immobility, as well as general balance issues, which can be overcome with adequate stretching, mobility, and vestibular training.

Test 5: Lunging

Purpose

To assess the asymmetrical movement of lunging. Many daily activities require use of the lunge pattern and its variations and, for this reason, the following tests look at the kneeling lunge, the dynamic lunge, and stepping.

Action

Kneeling lunge: standing, the subject steps forwards and kneels on one leg. This position is held for up to 10 seconds, as the practitioner observes joint alignment and overall balance. The test is repeated on the other side.

Fig 4.16 Observation of the kneeling lunge

Dynamic lunge: the subject stands about 15–20cm away from the centre of an open doorway. They are instructed to lunge forward over the centre of the doorway. Once the lunge is made, they should return to the starting position. This may be repeated several times before repeating on the other leg. The practitioner observes joint alignment and overall balance during the movement.

Fig 4.17 Observation of the dynamic lunge

Stepping: the subject stands facing a 15cm step. They step onto it and bring their feet together to stand. This may be repeated several times before repeating on the other leg. The practitioner observes joint alignment and overall balance during the movement.

Observation

The kneeling lunge is a pattern that is used regularly, especially where an individual has to bend down and remain in that position for a period of time. As the subject lunges forwards, there should be little disturbance to overall balance, with the maintenance of optimal spinal alignment; the subject should not appear to 'fall' into the movement, but exhibit good muscle control in the lead leg. The knee of the lead leg should remain in line with the second toe (longitudinal axis of the foot).

If balance is compromised, the subject may reach out for the floor or other suitable object to stabilise themselves.

Fig 4.18 Observation of stepping

Fig 4.19 An unstable lunge

Poor distribution of weight between the front and back leg may result in premature muscular fatigue, which can subsequently affect the ability to rise from this position without support. Maintenance of optimal spinal alignment requires adequate flexibility in the hip flexors (rectus femoris and psoas) of the rear leg; shortness in these muscles may result in excessive hip flexion, pulling the trunk forwards and causing instability. Another strategy that is often employed (and taught) is activation of the abdominal wall to produce stability in this position artificially. While this will create stability, improving flexibility in short muscles and increasing awareness of weight distribution can achieve the same objective with less effort.

During the dynamic lunge, there is no off-loading of body weight, and so the movement requires more stability and muscular endurance. As with the kneeling lunge, the trunk should remain well aligned over the pelvis throughout, with the knee joint tracking over the second toe. There will also be a visible push of the lead leg into the floor during the return, which helps to accelerate the centre of gravity backwards over the rear leg.

A common observation is seen where the knee moves inwards; this can not only contribute to instability, but also medial knee pain (chronic). In this case, the lateral hip rotators may be weak, or poorly controlled.

Shortness in the rectus femoris of the rear leg can restrict the stride length of the lunge, and shortness in the hamstrings of the lead leg can contribute to trunk flexion as the low back flattens. Improving the flexibility in these muscles can dramatically improve postural control. While these variations may exist, the practitioner should not overlook how the subject positions their feet during the lunge; the closer the feet get to the line of propulsion (the in-line position), greater stability will be required. Re-education of proper foot placement can go a long way, even in the presence of marked muscle shortness/tightness.

The practitioner should also consider observations of the lunge in different directions (if appropriate), for example, the side lunge and backward lunge.

Fig 4.20 Medial hip rotation when lunging

During stepping, the lumbar spine should retain its lordotic curvature, without excessively flattening (or arching). The knee joint should track in line with the second toe throughout the movement.

Fig 4.21 Optimal tracking of the knee when stepping

The point at which the rear leg leaves the floor indicates a position of instability; to preserve balance effectively, the abductor muscles of the stepping leg must contract to maintain a level pelvis (when viewed horizontally). If these muscles are weak or have poor control, the pelvis (on the side of the stepping leg) will move laterally and elevate (hip adduction), and the opposing side of the pelvis will drop below horizontal (hip abduction). This is often referred to as the Trendelenburg Sign. Weakness in the abductors may also be observed during standing/supine posture where the hip hikes on one side. Functionally, it can result in low back pain as the low back muscles overwork to stabilise the pelvis; knee pain may also occur because the hip joint becomes misaligned in relation to the knee joint. While a small amount of weight shifting towards the stepping leg is required during the movement, excessive shifting may result in a loss of balance. The return to the floor should be observed as the reversed pattern.

The stepping movement can also be observed as a forward step down, and may provide more information about pain patterns when going down the stairs. As well as observations of joint alignment and balance, the practitioner should also consider the differences in the type of muscle contraction when comparing stepping up to stepping down.

Test 6: Walking gait

Purpose

To assess the level of integration of the upper and lower extremities during walking. The evaluation of walking is one of the most useful assessments that can be made of postural control. As upright humans, walking represents a unique movement pattern, yet it is often performed with remarkable inefficiency.

Action

The subject is asked to walk at a comfortable pace. The practitioner observes joint alignment, sequencing of body parts, ease of movement and overall stability.

Observation

The assessment of walking should begin with an observation of the entire body, particularly at the initiation of each step. Efficient use of the body in walking will almost always begin with a slight forward shift of the centre of gravity that is initiated in the upper body. Almost simultaneously, there will be a slight forward motion at the knee of the lead leg, which passively flexes the hip ready for the first step. The forward momentum of the upper body over the legs will require very little effort by the lead leg, to perform a 'controlled fall' into the first step. From here on, the motion is cycled from left to right.

It is common for many people to flex the hip to lift the leg, before placing it down. People with short hip flexors often have increased knee flexion during walking, which may be a long-term risk factor for patellar tendinopathy.

Each knee joint should track over the longitudinal axis of the foot; excessive deviation medially and laterally may be associated with ankle pronation and supination patterns respectively (which may be observed during a

Fig 4.23 Good use of the body when walking

standing assessment of posture). As the foot moves towards heel strike, a slight lateral rotation of the hip will allow the gluteal muscles to engage upon weight bearing. This is assisted by the transference of weight to the outside edge of the heel – supination – which also helps to create stability in the ankle. This is quickly followed by a movement towards pronation, as full body weight is transferred to one leg, finally resulting in toe-off as the foot re-supinates.

Shortness in the calves may affect walking by limiting dorsiflexion, resulting in reduced heel strike; the knee may flex to compensate for this, as it reduces the tension on the calves. While shortness in the hamstrings will not affect stride length in walking, the forward displacement of the hips will contribute to poor efficiency.

Upper body observations include relaxed counter-swinging of the arms and a small amount of counter rotation in the thoracic spine and shoulders. This mechanism is important not only for maintaining balance, but also to help dissipate compression forces through the spine.

Finally, the head should remain above the shoulders and there should be little visible tension in the neck muscles.

A MODEL OF RE-EDUCATION

5

Introduction

The model of re-education used throughout this book is inherently simple and uses a number of effective movement-based techniques. It contains a multitude of tools that are not new, but the way in which they are combined may offer something new and useful, with the objective of increasing postural awareness and control at every stage.

Increased postural awareness leads to increased control of muscles and movement. This objective can be met quickly, particularly if the individual already possesses good body awareness and is pain-free. However, if the individual has poor awareness and/or chronic pain, additional techniques may be employed to help release or condition muscles and movements prior to and in conjunction with postural awareness techniques.

The outcomes of increasing postural awareness range from injury rehabilitation, through to general conditioning and sports performance. Increasing postural awareness can be achieved by diminishing the effects of gravity. It has been suggested that up to 85% of all central nervous system (CNS) activity is accounted for by the body's reaction to gravity, including the vast number of proprioceptive mechanisms responsible for keeping us upright and balanced. At first glance, this may appear to be a large percentage, but when you consider the anti-gravity role of the extensor muscles, coupled with the reflexive control mechanisms of the vestibular system, it soon becomes clear that this level of neurological input is required. With such a large proportion of the nervous system activity already accounted for, it leaves very little for the purpose of learning.

By diminishing the effect of gravity on the kinetic chain, we have a range of resources that can further enhance any type of learning, including awareness of posture, both statically and dynamically. When such a situation is created, muscles do not have to work directly against gravity, and the vestibular system will have lower demands placed on it. The cumulative effect is a system that is more self-aware and more able to learn, especially if the practitioner is able to structure movements and situations that are seen as new and challenging to the individual.

The simplest way to achieve a dramatic reduction in the proprioceptive background due to gravity is to lie down in a horizontal position: supine, prone, or side-lying. In this position, the practitioner will be able to 'listen' effectively to the individual's body and body parts as muscles are released and the limbs are guided through movement. In this position, the individual will be able to direct their attention more fully to any feedback, which will in turn lead to longer-term learning. This learning can be guided by the practitioner's hands or words alone, or by a combination of both.

Where muscle tightness exists, it should immediately begin to diminish in supine/prone/side-lying positions; if it continues to occur, then the practitioner can use muscle release or contract-relax techniques that may help to reduce tone further, and improve freedom of movement. Once this has been achieved, restrictions and limitations may be lifted sufficiently for new movement to take place: this prepares the body for larger, more complex movement patterns that can be continued in the lying positions (assisted or unassisted), or in other positions that progressively bring the individual back in touch with gravity.

This chapter aims to introduce the practitioner to a model of movement-based postural re-education. The use of release techniques as effective methods in preparing the body for change will be discussed, and the key processes of integration, feedback and exploration will be rationalised as potent sources of learning.

Increasing freedom of movement

All too often, muscles are passively stretched or massaged excessively to obtain 'release'; these types of approaches may bring about a perceived reduction in tonus (due to endorphin release), and may even temporarily reduce tone, but may not offer a long-term solution. Prolonged use of these techniques may also result in unwanted structural changes in the muscle and joint structures that affect function negatively. By teaching the muscles how to relax instead, by a reduction in neural input, a new pattern can be learnt and maintained unconsciously; improvements in muscle control can be re-established and the released muscle can then be fully integrated into new or existing movement patterns, with a resulting improvement in postural awareness.

Although many different forms of achieving muscle release and freedom of movement exist, two simple formats are discussed below: positional release and contract-relax.

Positional muscle release

Although there are over 20 distinct definitions, 'release' can be thought of as a 'freeing-up'

process. In the context of postural re-education, it may be interpreted as a reduction in muscle tonus, for the purpose of reducing pain and/or increasing freedom of movement. The principles of positional release can be best explained with the following metaphor: imagine a rope that has a knot in the middle. If the rope is pulled apart (stretched), the knot becomes tighter and harder. However, if we bring the two ends of the rope together, the knot begins to release and become looser.

If a muscle is shortened through contraction, the body can be passively positioned to shorten the muscle a little further (i.e. the origin and insertion brought closer together). This passive positioning allows two things to take place. Firstly, it takes the strain of a contracted muscle by allowing it to complete the motion of contracting. This in turn sets off a reflexive response within the body to begin adjusting the position of the opposing muscles to accommodate this change in muscle length. Secondly, there is a neurological response through the proprioceptive reflexes that induces a relaxation in the muscle (reduction in tonus). Once a release is found, joint compression may also be applied as a way of shortening the muscle further. A release position is normally held for up to two minutes, and there may be a palpable softening of the associated muscle(s). If there is appropriate feedback from the practitioner, this will further clarify the information coming into the proprioceptors, and the body begins to understand how to be more comfortable.

The main purpose of this type of release is to begin a process of relaxation of the involuntarily contracted muscles. Once muscles have been released in this way, it is often useful to begin a process of 'integration' whereby muscles are integrated into larger movement patterns – a process that may be continued in both a lying or upright position. In this way, the practitioner takes the individual through a series of movements which introduces them to more efficient joint alignment and movements that were previously unavailable due to tight contracted muscles. Because the individual is often very relaxed by this point, they are more able to interpret fully the guided movements of the practitioner. This type of education may be achieved through verbally guided movement sequences or by practitioner-assisted movements. If space permits, then a combination of hands-on manipulation and verbally guided movement is often the best approach. Some of the lesson plans in Part Two make use of positional release early in the sequence, to maximise movement potential.

Contract-relax

Contract-relax techniques, including proprioceptive neuromuscular facilitation (PNF), have been traditionally used to increase passive flexibility of a muscle. While these methods do not directly influence muscle tone compared to positional release techniques, they override the stretch reflex for a small period of time, during which muscle tension is reduced. This period of inhibition allows the muscle to be stretched further with considerable ease, and the stretch receptors of the muscle spindle immediately accommodate a greater muscle length. In its simplest format, the mechanism works as follows: after assuming an initial passive stretch, the muscle being stretched is isometrically contracted anywhere from 5–15 seconds, after which the muscle is briefly relaxed for 2–3 seconds, and then immediately subjected to a second passive stretch. This stretches the muscle even further than the initial passive stretch, and is normally held for 5–15 seconds. The cycle can then be repeated as many times as is necessary in order to achieve the desired change. The muscle is usually relaxed for 20 seconds following this procedure.

In situations where muscle hypertonicity and muscle shortness are present, a release closely followed by a contract-relax technique may provide a very quick improvement in freedom of movement around a joint. The newly relaxed muscle may now stop inhibiting its functional antagonist, as well as possessing improved flexibility. From here, the muscle and associated joints can be re-introduced to movement without restriction or limitation.

Many of these techniques can be self-administered or practitioner-assisted, as well as verbally guided. During assisted techniques, the practitioner will guide the subject through a series of hands-on controlled muscle contractions (and subsequent releases). Many of these releases may be consolidated further by engaging other associated muscles.

Integration

The process of integration constitutes an important step in postural re-education, and should not be considered to be a final step but as an on-going process. Once muscles have learned how to release, the potential for movement can be realised and the stage is set for further learning. For a full learning experience to take place, it is important that movements are introduced immediately after release techniques to continue the 'awareness conversation' with the body. The movements are performed in a slow, gentle manner, and should be explorative rather than goal-orientated; many of them are performed in lying positions to diminish the effects of gravity, although further integration should be pursued in upright positions and using load where necessary.

These sequences usually aim to combine several movements which are thematically organised around a functional action, such as the developmental movements of rolling, crawling, and standing up; clinical practitioners may teach their patients movements based around reducing muscle tone, improving joint range of motion, or improving breathing; fitness professionals may advocate movements that involve systematic explorations of the kinetic possibilities of joint and muscle groups; athletes and performing artists may wish to experiment with movement-based imagery and visualisations that relate to perception, cognition and other aspects of motor function.

A multi-modal approach

As previously mentioned, these sequences of movement are often preceded by (and combined with) hands-on manipulations, such as release techniques. However, the many formats of manual therapy and exercise training should also be viewed as potential aids. For example, careful and intelligent administration of massage, chiropractic, osteopathic, stretching and strength training techniques can play an important part in overall postural awareness and re-education. Regardless of the techniques used, the prime objective is to set up a dialogue between the individual and the practitioner that is conducive to learning. This is dependent upon the creation of an environment in which information can be effectively delivered to the individual, as well as feedback received from the practitioner.

Other important considerations

Testing and re-testing

Testing and re-testing are useful feedback tools which bring the body into sharp awareness with itself, by increasing the conscious connection with existing and new posture. It can also reveal how body parts move in integrated ways and feel coordinated with each other. In this way, the subject is made conscious of what's 'switched on' and what's 'switched off' at the same time. Now the body becomes aware that it is easy to get messages through to a body part that was previously kinaesthetically unaware. Subjects who already possess good body awareness can feel the improvements almost immediately, and often report an enhanced sense of connectivity to their body. The effectiveness of many of the lessons in this book lies in the process of re-testing; this in effect consolidates the new pattern and subsequently modifies the original motor programme, which is then stored in the brain. These new programmes will gradually become the body's automatic and preferred response to movement. This kinaesthetic conversation is an aspect of postural re-education that is fast, thorough and painless, as it does not require large amounts of force.

Higher integration

After muscles have been released and re-educated through movement, an individual's own daily activities can continue to strengthen their previously inhibited or weak muscles, providing these activities allow for continual and varied stimulation. This is based on an assumption that the individual does not return to the faulty movement patterns that caused the problems in the first place. Therefore, further lessons in alignment and posture that incorporate any necessary functional demands, such as loading, may be necessary.

In the context of athletic performance, movement patterns may also need to be modified (as they are now being performed with increased freedom of movement), usually by the use of sports-specific drills. This will help to anchor the re-education process within the sport, and create lasting changes.

Sometimes the integration of new information that has been introduced in a single session continues to happen over a period of days; changes in awareness and movement will continue to evolve and the individual will have regained more functionality by the next visit.

Process of exploration

Optimal learning involves a process of exploration. For example, during the crawling stage, a baby does not yet possess the musculoskeletal and neuromuscular adaptations that allow it to stand upright; yet they continue to move about, playfully exploring their immediate environment, without necessarily being overly concerned about what they are doing. As they move around and meet new stimuli, small children will often stop and proprioceptively absorb the new stimuli, before continuing. The new proprioceptive input is easily absorbed (and learned) because the child is learning how to move, rather than focusing on an end goal. Being able to pause at any time during the movement before continuing from exactly the same point (a phenomenon also reflected in animal behaviour) further supports the notion that there must be a high level of (unconscious) proprioceptive control during the process of movement. Reversibility of movement is a concept that can quickly enhance postural awareness and subsequent control, and will be explored in the next chapter.

Process vs. content

As adults, we often tend to move in ways that focus purely on what we are doing, rather than how we are doing it. For example, focusing on turning our heads to see something, as opposed to focusing on how we can turn our heads; we focus on the content of the movement, and not the process. As our movements become more and more content-orientated, we gradually lose the ability to notice the process of moving, and consequently lose the awareness of the movement; with this loss comes an inability to change the movement at an unconscious level. By learning how to shift our conscious attention temporarily to how we are performing a movement (i.e. the process of doing it), we suddenly become aware of how we can begin to change

it. Over time, these changes can then become ingrained unconsciously once again, requiring less and less conscious intervention. This in turn will not only lead to longer-lasting changes in posture, but will also make habitual posture more energy efficient, as we learn to use the least amount of effort to move.

Eliciting the process

During movement lessons, the practitioner should aim to focus on guiding the individual verbally (and sometimes physically) through the process of moving, rather than the achievement of a pre-defined goal. This may be done in several ways but in simple terms involves a few carefully directed questions. Below are some examples of process-orientated questions that can be asked during movement:

- How does the movement feel to you?
- How else can you do this movement?
- Where is your attention as you move?
- How does the movement compare to your other side?
- What happens if you do it this way?
- What do you feel working?
- How does the movement feel freer?

The above questions are useful in that they compel the individual to increase their awareness so they can answer them, thus bypassing the end goal and facilitating learning. When the individual performs a similar pattern of movement in daily life, they can then begin to draw on this learning, which will enable them to move more efficiently. Good questioning should provoke an almost child-like sense of wonder in the individual, as if learning is a new and exciting process. Because optimal learning usually takes place as a result of trial and error, the practitioner should not attempt to lead the individual in any way, but rather invoke a sense of exploration. Sometimes, a completely different movement or position may be required to re-establish the learning process, and for this reason, there is no one technique that will work alone for every subject. For this reason alone, it's important to teach many different movement patterns in many different ways to increase the options available.

Hands-on awareness

Finally, practitioners should feel confident in guiding individuals through movement physically. This can be extremely effective, especially when postural awareness is poor. For example, many people find the achievement of good head carriage to be difficult through verbal instruction alone; gentle navigation of the head (and the rest of the body) into a more favourable alignment will speed up the process of re-education. Similarly, during more complex sequenced movement patterns, the gentle nudging of body parts can improve efficiency by enhanced proprioception and control. The use of the hands is a skill that requires good knowledge of simple biomechanical principles, as well as confidence to apply these skills; above all, it demands a keen eye for spotting good movement, as well as compensatory pattern.

PART **TWO**

EXERCISES FOR POSTURAL TRAINING

STRATEGIES FOR POSTURAL RE-EDUCATION

6

Introduction

The maintenance of posture is under unconscious control: the reason that most of us can assume an upright position reasonably comfortably is not a product of conscious control. At times we may become directly aware of our posture, but this is usually when it causes dysfunction or discomfort; suddenly we realise that we are not aligned or stable, and momentarily our nervous system shifts our attention to it, as we re-adjust appropriately. As postural disturbances become more frequent, we can lose the ability to adjust effectively, until at some point, we can have almost complete 'amnesia' for what is optimal body configuration. At this point, faulty alignment is now perceived as normal and subsequently becomes an unconscious pattern, which is repeated, causing longer-term dysfunction and impairment. Changing such a perception, whether to reduce pain or to prevent pain and maintain good health, is an important result of postural care programmes. In the majority of cases, this process relies heavily on improving body awareness, which provides an effective point of reference for subsequent re-wiring or adjustment of the body. The results of postural re-education are:

- To provide the body with sufficient motivation to begin to change its own configuration to one of the least stress and restriction.
- To teach the body how to increase awareness of posture, so that it learns to make distinctions between optimal and less than optimal posture, as well as learning how to return to

postural equilibrium quickly, following a perturbation.
- To find ways of installing these strategies unconsciously so that the nervous system can use them without conscious effort.

This chapter explores how these results can be achieved by drawing on a diverse range of methods. The success of these methods is not necessarily dependent on the skilful application of individual techniques but rather the way in which the techniques are strategically sequenced and progressed. Many of these progressions have their roots in developmental motor patterns, with a strong emphasis on function. The majority of them also involve no more than the simple manipulation of body weight in space. For ease of use and application, these methods of re-education have been woven into individual lesson plans that contain specific movement sequences, thematically organised around common postural objectives. The aim of all the lesson plans is improved control and quality of movement. If and when further performance enhancement is warranted, these guiding principles can be applied while further manipulating acute exercise variables, including sets, repetitions, load, tempo and rest periods.

Strategic elements

The use of the term 'lesson' is important because it allows the practitioner to move individuals towards a process of learning, rather than simply doing. In turn, the individual will learn to regard postural movement as a learning experience. To aid this process, a number of strategic elements have been identified to maximise learning and change, and the practitioner should incorporate these principles as much as possible when selecting and delivering the lessons.

Increase body awareness

In its simplest sense, an individual's awareness can be increased by directing them towards their own perception of posture as it currently exists; if their posture is then modified in any way, they will have a reference point, which will feedback information about how their posture has changed. Awareness can take many forms for different people, such as changes in muscle tension or joint alignment; there may even be a sense that something 'doesn't feel right'. The practitioner should look for ways to maintain continual awareness of postural change, and in a way that the individual prefers, as well as introducing other ways of feedback.

Improve transitional control of movement

The sequencing of different movements can require large amounts of proprioceptive and exteroceptive input, which can greatly enhance overall postural control. It places a number of simultaneous demands on the individual including awareness of the centre of gravity and joint alignment, spatial awareness, and appropriate application of ground reaction forces. Postural control may be further challenged via performance of the sequences in reverse.

Use over-correction

Over-correction principles can be used to accelerate learning through manipulation of acute exercise variables. For example, where control

of balance is poor, use of an unstable surface can help to engage co-contraction (reciprocal activation) mechanisms that might otherwise not be available on a more stable surface. Over-correcting balance in this way will help to provide the individual with a frame of reference that is further away from their existing reference. What immediately follows is a recalibration of balance towards the new frame of reference. It is important that the practitioner ensures the individual is ready for such a progression, and it is equally important to understand that over-correction is just a stepping-stone for progress.

Create a diversity of options

The practitioner should always devise challenging situations and movements to encourage diversity in postural control. With more ways of controlling posture in many different situations, the body will always be able to respond to potential disturbances.

Install motivation

Within the framework of posture, it is important to create the right type of motivation to elicit change. For example, if we consider the motivation behind a squat to stand movement from a childhood developmental perspective, it is simply to be able to obtain a higher vantage point – to be able to explore the world from a higher position. If this strategy is now incorporated into a series of squats by asking the individual to imagine they are trying to stand up as if looking out of a high window, the movement experience will be vastly different to one where no motivation is given.

Encourage feedback

For any postural change to occur, feedback is important. This feedback may be extrinsic –

71

from the practitioner, friends and family; or it may be intrinsic – the individual may experience less pain or better balance. Because the vast majority of postural control is heavily reliant on intrinsic feedback, it should be encouraged in all stages of learning. As the individual achieves better control of posture, these internal feedback loops will become more tightly regulated.

Encourage minimal effort

Throughout all postural lessons, the practitioner should continually encourage movement that uses minimal effort. While initially this may be challenging, if regularly reminded, the individual will learn to make fuller use of reciprocal inhibition and less use of reciprocal activation as a way of preserving economy of effort. Where excessive effort is used, the practitioner can substitute effort of movement by appropriate support or application of force, as an aid to learning.

Encourage exploration

Where possible, individuals should be encouraged to explore movements themselves during postural lessons. This process helps rapid learning by allowing individuals to maintain a continual feedback loop resulting in flexible motor behaviour.

Lesson guidelines and instructions

It's worth noting that many of the individual movements within the lessons may already be familiar in other guises, having been appropriately modified to fit within the framework of postural re-education. Particular attention should be paid to these subtle changes, as many of them represent significant shifts in awareness and perception that are often essential for postural control.

Each lesson makes use of sequences of movements that help to focus awareness on how the body organises itself in preparation for, and during, movement. Some of them may seem unusual in that they are not familiar; however, if the focus of attention is put on the process of movement, rather than the end goal, then the outcome will be improved postural control, both statically and dynamically. Patience, deliberation, exploration and effortless movement are all virtues to be encouraged when taking part in these lessons.

It is important to understand that the exercises and movements in this chapter are in no way exhaustive; however, they have all been tried and tested by myself on thousands of individuals from a variety of backgrounds and objectives. They represent a collaboration of thoughts, ideas, paradigms, techniques and anecdotes from educators past and present. What may be regarded as different is the way in which the techniques are brought together as strategies – a process that is central to this model of postural re-education. With this in mind, the sequences should be adhered to in the first instance; however, as the practitioner begins to understand how these movements affect posture, individual movements may be woven together in different ways and with new movements, to produce original lesson plans. I request that every practitioner explore this avenue so that newer ways of re-educating posture can be discovered, through a process of trial and error.

The following guidelines will enhance both the sense of awareness that these movements bring and the outcome of improving postural control.

Clothing and footwear

Clothing should be comfortable and loose enough to allow full range of movement. Because many of the lessons involve movement on the floor and across other surfaces such as stability balls, appropriate clothing should be chosen that will not cause chafing against the skin. The majority of the movements should be performed in bare feet, if comfortable.

Prepare a comfortable space

Movements are best performed on a surface that is comfortable for bone contact, yet firm enough to clearly provide feedback for movement. There should be enough room to slide the arms and legs in all directions.

Mindful repetition

The vast majority of the movements are not designed to be forceful. They should be performed until a sense of change in the movement becomes apparent; for example, it begins to feel relaxed or more coordinated. If a movement feels particularly challenging, ways should be found to make it easier.

Maintain a feedback loop

During and after each lesson, aspects of posture will begin to change. This may include feelings of relaxation, lightness, warmth or simple feelings of being more connected to various body parts. Sometimes this may not be immediately apparent but will be perceived as feeling different. A return to standing posture is often encouraged so that any new and unfamiliar sensations can become integrated with existing posture.

How much and how often

There are two ways in which the lesson plans may be used to improve postural control.

Firstly, relevant lessons may be used where specific postural objectives need to be met. Secondly, the lesson may be used more generally to enhance existing postural control on many different levels. As postural control becomes more efficient with time, the frequency of the lessons may decrease, and more emphasis should be placed on the use of more complex and diverse movements that challenge postural control. It is important to encourage movement outside the clinic or studio. This may be achieved by home movement plans and more importantly, by integration of learning into activities of daily living.

Health notice

All the movements outlined in this section demand levels of cardiovascular fitness, strength, flexibility and balance that are similar to a beginner's exercise class, and are suitable for almost all ages and abilities. If you are in any doubt as to whether they are suitable for you, please seek the guidance of a qualified health professional. If you have any medical problems or conditions that might be aggravated by movement or exercise, you should consult your GP before doing these lessons.

While every effort has been made to include comprehensive details of each lesson, some of the movements may be ambiguous, and may be subject to misinterpretation; however, you should experience no pain or discomfort if the lessons are done correctly. As with any learning experience, the written word alone is no substitute for individual instruction and demonstration, where the subtleties and nuances of the movements can be explored more fully and clearly. Anyone wishing to do these lessons as they were originally intended should seek the guidance of a suitably qualified specialist, in conjunction with using this book.

LESSON PLANS FOR THE LOWER BODY

7

Lesson 1: Increasing awareness of the foot muscles

Theme(s)

A series of simple bare foot movements that explore how the muscles of the plantar surfaces of the feet control movement and provide support; the intimate relationship between the toes, arches and ankle is also explored. This sequence may also be useful as a foot refresher, in preparation for movement.

Postural benefits

Restores and maintains proprioceptive control of the foot; improved balance in standing; increased strength of the toe flexors.

Movement sequence

- Passive toe stretch
- Active toe stretch
- Toe mobility
- Foot shortening

1. Passive toe stretch

The toes are passively flexed and extended holding each for 20–30 seconds. A lateral stretch can also be performed by placing your fingers between each toe to spread them gently.

The arrow denotes the pulling of the toes.

Fig 7.1 Passive toe stretch

2. Active toe stretch

The toes are actively flexed, extended and spread, holding each position comfortably for 20–30 seconds.

The arrow denotes the active pulling of the toes towards the shin.

Fig 7.2 Active toe stretch

3. Toe mobility

Flexibility can be integrated into simple mobility movements such as toe wiggling and toe waves. Movements are randomly maintained for 60 seconds, with emphasis on movement of all the phalangeal joints.

The arrow denotes the curling of the toes underneath the foot.

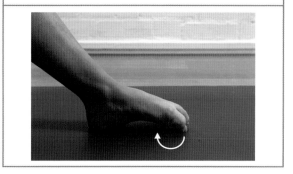

Fig 7.3 Toe curl

4. Foot shortening

Control of the toe flexors can be enhanced further by placing your bare foot on a flat surface and performing toe curls and toe drags against the resistance of the surface, while keeping the heel fixed.

The arrow denotes the dragging of the toes backwards towards the heel.

Both movements will reveal a different pattern of muscle recruitment, and emphasis should be placed on performing slow contractions and releases for about 1–2 minutes. The toes should be actively or passively extended following these movements to ease any stiffness or discomfort.

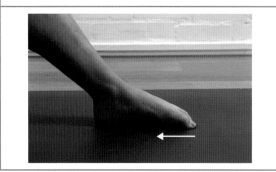

Fig 7.4 Toe drag

Lesson 2: Restoring neutral alignment of the ankle

Theme(s)

A sequence of progressive movements that help to restore control to the muscles responsible for maintaining arch support in the feet. The movements are best performed in bare feet that have been warmed up with the movements from lesson 1.

Postural benefits

Restoration of neutral ankle alignment in those with marked pronation; improved strength and control of the tibialis posterior muscle.

Movement sequence

- Freeing up the ankle
- Neutral ankle awareness
- Muscle awareness
- Improving control of the tibialis posterior

---

1. Freeing up the ankle

Begin in a seated position with your feet flat on floor, hands on your thighs. Slowly explore the movement of lifting your toes upwards, followed by lifting the ball of your foot off the floor, until you are resting your weight on your heel. Slowly repeat six times.

The arrow denotes the lifting of the balls of the feet upwards.

Fig 7.5 Toe lift

Notice how your knee, hip, pelvis and low back all move slightly. Now explore a movement of lifting your heel off the floor, noticing how your knee, hip, pelvis and back are moving now. Repeat six times and rest. Now combine both movements slowly, for a total of six cycles – notice how your knee, hip, pelvis and back move rhythmically with your ankle. As you rest, notice how your foot is sitting on the floor.

The arrow denotes lifting of the heels upwards.

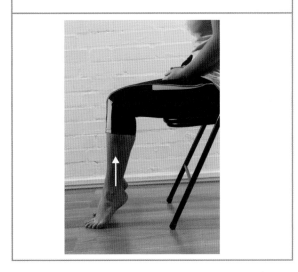

Fig 7.6 Heel lift

With the foot flat, begin to explore the movement of rolling your foot to its outside edge and back again. As you do this, allow your body weight to shift gently over the same side hip, so that you slowly sway with the movement. Notice how your knee moves out a little. Repeat six times. Perform the same movement but rolling to the inside edge of your foot, noticing the opposite motion of your knee and body sway. Pay attention to all these movements as you perform six repetitions. Now combine these two movements for a total of six cycles. How does the foot feel now?

The arrow denotes rolling to the outside edge of the foot.

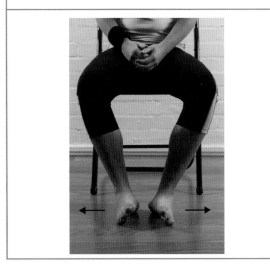

Fig 7.7 Lateral roll

2. Neutral ankle awareness

In a seated position, with your foot relaxed, notice the shape of the arch. Now, keeping the ball of your foot firmly on the floor, lift your toes upwards (extension) and hold this position for a few moments. Notice how the arch is lifted higher, putting your ankle in a neutral position – you may even notice the knee moving laterally.

The arrow shows lifting of the arch of the foot as the toes lift.

This is known as the Windlass Effect. As you become more aware of your ankle in neutral alignment, practise retaining this alignment as you relax the toes. When you can hold your ankle in neutral for 60 seconds or more, you can progress to doing this movement when standing. The addition of bodyweight to this movement will begin to bring an awareness of the muscle action required to hold this position, which is explored further in the next step.

3. Muscle awareness

In a seated position with your feet on the floor, perform the movement in step 2 to put the ankle in neutral. As you release your toes, maintain the arch support and notice if you are aware of any muscle activity in the calf. You may be aware of a slight contraction on the medial side of your calf. If you cannot feel this contraction, try palpating the area or imagine lifting the apex of the arch (as if inverting your foot slightly). This muscle action is the tibialis posterior. Practise these muscle contractions methodically, with initial focus on simply engaging it, noticing how it contributes to lifting the arch. For additional proprioceptive integration, you can gently contract the intrinsic foot muscles by performing a toe drag (see lesson 1) as you contract the tibialis posterior.

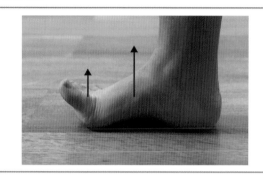

Fig 7.8 Neutral ankle via the Windlass Effect

The movement begins by gripping the floor with the toes for about five seconds – you will feel the muscles of your foot and calf contracting. Allow the contraction to ease off smoothly over five seconds, and as you do so, allow your heel to drop about 1cm as your toes relax.

The arrow shows the toes curling inwards.

Hold this new position, and repeat the contract–relax cycle until the heel touches down, and the entire foot is contacting the floor. When the heel touches down, wiggle the toes of that foot for 10 seconds. Slowly relax the foot and walk around, noticing the difference. Repeat on the other side.

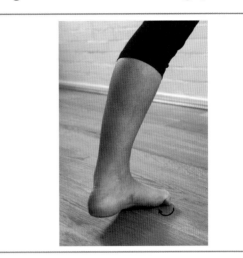

Fig 7.11 Calf stretch – toe grip

2. Tibialis anterior stretch

The subject lies supine with their hips flexed to 90°, knees bent, feet standing. The practitioner places one hand on the top of the foot and the other on the knee for support. The subject dorsiflexes their ankle, engaging the tibialis anterior muscle. The practitioner matches the upward pull of the ankle and resists movement for five seconds.

The arrow on the trainer's hands shows the direction of force. The arrow on the subject's foot shows the direction of pull upwards.

The subject slowly releases the contraction, and immediately pushes the ball of their foot into the surface. This is held for five seconds before relaxing, at which time the practitioner should be able to move the foot further away from the subject. The contract–relax cycle is repeated three more times.

3. Freeing up the ankle
Refer to lesson 2, step 1

Fig 7.12 Tibialis anterior stretch

4. Heel/toe/edge walks

These three movements are performed for 1–2 minutes each. The heel walk involves walking on the heels; toe walking involves walking on the toes; and edge walking involves walking on the lateral edge of the feet. The movements should be performed on a floor that is suitably padded to prevent discomfort.

5. Single leg balance – stable surface

The single leg balance is an important progression for enhancing postural control. A number of variations exist that provide challenges in all three planes.

The first variation is the sway: while balancing on one leg, maintain good postural control as you begin to sway gently forwards and backwards. A sway may also be performed in the transverse plane as the body gently pivots around the hip (and ankle). Frontal plane sways are likely to be the most challenging, only offering a very small amount of movement. It is important not to sway so far that you have to step out of the movement. Pay attention to how your body organises itself when swaying; this is an important mechanism for postural control and should not be inhibited, especially in favour of abdominal wall recruitment.

Single leg balance may be further challenged by the movement of other limbs. For example, moving the opposite hip through all its possible movements; moving the spine through flexion, extension, lateral flexion and rotation; or exploring movements of the shoulders, head and eyes. Notice how these movements affect your balance, and how other body parts re-position themselves to support balance. Practise all these variations for 30–60 seconds each.

Fig 7.13 Single leg balance sway

Fig 7.14 Single leg balance with hip extension and arm movements

For a really challenging approach to improving control of the ankle (and overall balance), perform clock touches as follows: stand on your left leg and imagine you are standing in the middle of a clock face, with 12 o'clock in front of you and 6 o'clock behind you. Now, using your left hand, slowly reach forwards towards 12 o'clock, and reach the right foot backwards towards 6 o'clock. Hold for a moment, before returning. Repeat this for different number combinations.

Fig 7.15 Clock touches

6. Double/single leg balance – unstable surface

The use of unstable surfaces will help to create 'safe emergencies' for the body that speed up our natural homeostatic mechanisms to postural disturbances, such as momentary losses of balance. A good starting point is to stand on a piece of foam or walk across a crash mat to challenge balance on two legs; further progressions can then be made by standing on one leg on a similar surface. For more advanced progressions, use an inflatable balance disc. Practise double and single leg standing for up to 60 seconds.

Other variations and progressions include the use of rocker boards, balance discs and wobble boards, all of which will increase somatosensory awareness in the sole of the foot, and thus integrate the action of the intrinsic foot muscles with those of the ankle.

Fig 7.16 Balancing on a disc

Lesson 4: Re-educating the gluteal muscles

Theme(s)

Using a variety of floor-based movements that engage the gluteal muscles through hip extension and abduction, before progressing through to upright patterns. This is a 10-step process that effectively builds on learning from each previous step, progressively increasing awareness and control of the gluteal muscles as important hip stabilisers.

Postural benefits

Increased stability of the hip joint; improved control and strength of the gluteal muscles; enhanced balance.

Movement sequence

- Abductor/adductor stretch
- Seated hip stretch
- Assisted hip release
- The clam
- Side lying abduction
- Prone squeeze
- Prone hip extension
- Modified shoulder bridge
- Kneeling squats
- Standing hip abduction/extension

1. Abductor/adductor stretch

To stretch the abductors, stand side on to a wall and step forwards of the inside leg. Place the inside arm on the wall for support and the other hand on the hip. Begin pressing the hip towards the wall and slightly downwards until you feel a stretch on the outside of the hip closest to the wall. Hold for 30 seconds, before repeating on the other side. Perform the stretch twice.

Fig 7.17 Standing abductor stretch, pushing in the right hip to the wall

To stretch the adductors, kneel on the floor and spread your knees as wide as is comfortable. As you breathe in for a count of five, gently rock forwards and squeeze your knees together into the floor. Exhale and slowly release the contraction as you sink further forwards. Come back to the start position, spread your knees a little further and repeat two more times.

The arrows denote squeezing the knees together.

Fig 7.18 Kneeling adductor stretch

2. Seated hip stretch

Sit on the floor with the front and back leg bent to 90°. Place one hand on the sole of your foot and the other on the floor beside you. Keep your pelvis tilted anteriorly and gently bend forward at the hip until you feel a stretch in your buttocks and hamstrings. As you inhale for a count of five, press your front knee down into the floor. Slowly exhale and release the effort, as you bend further forwards. From this new position, repeat two more times, stretching further each time.

Fig 7.19 Seated hip stretch

3. Assisted hip release

With the subject lying on the floor or couch, the practitioner holds one leg in 90° hip and knee flexion. The subject relaxes all efforts of holding their leg, as the practitioner slowly explores small hip movements (extension/flexion, abduction/adduction and rotation). As this is taking place, the practitioner 'listens out for' points of resistance in the movement. Each time the practitioner comes across such a point, the subject is asked to relax the effort of the muscles.

This process is repeated for 2–3 minutes or until all hip movements have become substituted by the practitioner and the joint action feels smooth.

Fig 7.20 Assisted hip release

4. The clam

Lying on your side, your knees and feet stacked vertically, with your knees bent to approximately 90°, your top hand is placed on the pelvis with your fingers spread across your buttocks. Keeping the pelvis fixed, lift the top knee away until a strong contraction is felt in the muscles underneath your fingers. This is held momentarily before slowly releasing.

Repeat until you can no longer perform a repetition with good movement (no more than 20 repetitions). Repeat on the other side.

Fig 7.21 The clam

5. Side lying abduction

Assume the same position as the clam but with your top leg extended. Keep your hand on your hip as before, and rotate your hip laterally to point your toes slightly upwards. Slowly raise your leg to just above horizontal and imagine you are reaching away with the leg.

Slowly return and repeat until you can no longer perform a repetition with good movement (or for no more than 20 repetitions). Repeat on the other side.

Fig 7.22 Side lying abduction

6. Prone squeeze

Lie on your front with your knees apart and bent to 90°. Place a small cushion or ball between your feet and slowly squeeze them together until you feel your buttocks contract. Ensure that you squeeze sufficiently enough to keep the rest of your body relaxed. Hold this contraction for three seconds before slowly releasing.

Perform 15 squeezes, focusing on the quality of contraction.

Fig 7.23 Prone squeeze

7. Prone hip extension

Following the prone squeeze movement, straighten both your legs. Focusing on one leg, begin slowly to take the weight off it until you feel your buttock contract. Hold for a few moments and slowly release. Repeat this movement, progressively moving your hip into further extension, until it lifts no more than 2–3 inches off the floor.

Perform no more than 10 repetitions, and maintain your awareness on the buttock contracting throughout. Repeat on the other leg.

Fig 7.24 Prone hip extension

8. Modified shoulder bridge

Lie on your back with your legs bent and feet flat on the floor. Position your heels just in front of your knees. Slowly push your feet down into the floor until you feel your buttocks contracting, then slowly release.

With each repetition, push your feet harder until your buttocks begin to lift off the floor, returning each time to the start position. As you progress with the movement, you will be able to lift your buttocks off the ground a little more each time as the spine peels off the floor, and the hips, shoulders and knees are in a straight line (bridge position). Aim to maintain a good buttock contraction throughout, for a total of 10 repetitions.

Fig 7.25 Modified shoulder bridge

9. Kneeling squats

In a kneeling position, place your knees about shoulder width apart, with your feet close together behind you. Imagine you are about to sit on a chair and slowly allow your hips to move towards your heels as your trunk bends forwards. Keep the spine in good alignment throughout, ensuring that your pelvis is tilted anteriorly. Squeeze your buttocks as you return to the upright position.

Perform 10 repetitions slowly.

Fig 7.26 Kneeling squats

10. Standing hip abduction/extension

The hip abduction movement is performed standing facing a wall with hands on the wall for support. Keeping both legs straight, with good spinal alignment, abduct the outer leg until a strong contraction is felt in the outer part of your hip. Slowly release and repeat for 10 repetitions, before switching legs.

Fig 7.27 Standing hip abduction

For the hip extension movement, stand facing the wall, with both hands holding it for support. Slowly extend one straight leg, allowing it to slightly abduct and laterally rotate. You should aim to extend it only as far as your pelvis and lumbar spine remain in neutral alignment. Repeat 10 times before switching legs.

Following this sequence of movements, notice how the gluteal muscles feel. Do you have more awareness of how they are involved in hip abduction and extension movements? Notice how they feel during larger hip extension movements such as squatting and lifting.

Fig 7.28 Standing hip extension

Lesson 5: Improving mobility of the hip joints

Theme(s)

A series of movements that enhance hip joint function by improving mobility and control.

Postural benefits

Increased control of range of motion in hip rotation; lumbo-pelvic-hip integration.

Movement sequence

- Hip release
- Supine hip opener
- Prone hip rotation (assisted)
- Seated hip refresher

1. Hip release

Refer to lesson 4, step 3.

2. Supine hip opener

Lie on your back with your legs straight and bend your left leg so that your foot is flat on the floor. Slowly allow your left knee to drop out towards the floor, ensuring that your pelvis remains fixed, and the movement only occurs at the hip joint. Start with just a few inches, gradually increasing the distance. Keep the movement smooth and coordinated, noticing how the foot rolls. Continue until you can no longer increase your range of motion without moving the pelvis. Repeat on the other side.

Fig 7.29 Supine hip opener

3. Prone hip rotation (assisted)

The subject is lying prone and the practitioner takes hold of their leg by the ankle and passively flexes the knee to 90°. This position is held for a few moments to allow the hip muscles to relax. Once the subject has begun to relax, the practitioner begins to explore medial and lateral rotation of the hip joint, by slowly moving the ankle laterally and medially, respectively. With each of these movements, notice how the pelvis is involved. At the limits of hip rotation, the pelvis is often required to participate to allow the movement to continue in that direction.

The arrow denotes the direction of pull.

Fig 7.30 Prone hip rotation

When the pelvis becomes prematurely involved alongside hip rotation, the effort of pelvic motion by the individual can be released if the practitioner places their hand under the pelvis to support it during the movement. In this way, the effort of pelvic movement can be effectively reduced, as the leg and pelvis now move as one unit. Often in this situation, the subject may 'feel no movement' at the hip joint and will be able to free up some of the restricted movement around the hip joint. As this happens, the subject may become aware once more of the possibility of movement involving the pelvis and leg, and be able to perform the original hip rotation with more ease.

The arrow denotes the direction of pull.

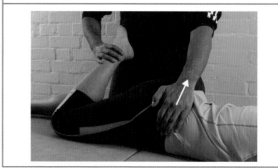

Fig 7.31 Prone hip rotation – substituting effort of pelvis

4. Seated hip refresher

Sit on a chair with your feet flat on the floor, and knees hip width part. Allow your spine to rest comfortably against the backrest, with your hands on your thighs. Place your left hand on the inside of your right knee, and your right hand on the outside of your right knee. Begin by gently pushing your right knee against your right hand. Push hard enough to allow your knee to move several centimetres outwards. Notice where you feel muscles contracting; if you are unsure push a little harder until you sense it. You will feel the right hip muscles but may also be aware of contractions in the right side of the abdominal wall and the right side chest.

Now, slowly relax your efforts and allow your right hand gently to push your knee back towards the centre, paying attention to those same muscles relaxing. As you come back to the centre, immediately push your knee into your left hand, allowing the knee to move inwards now. This time, notice how the right inner thigh muscles, the left side of the abdominal wall and the left side chest muscles are contracting. Once more, slowly relax your efforts as you return your leg to the centre, and repeat to the other side again. Continue with this movement for six repetitions in each direction, before repeating on the other leg.

Following this sequence of movements, stand up and walk around, noticing how your hip joints feel, and how this translates into your movement. Notice how much easier it is to rotate your trunk, and how less restricted the lumbo-pelvic-hip region feels.

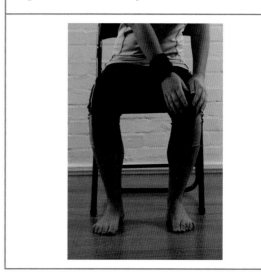

Fig 7.32 Seated hip refresher

Lesson 6: Learning how to bend

Theme(s)

Using inspiration from developmental movements to create progressive patterns that improve strength and control of the legs during bending. Many of these progressions challenge eccentric control of muscle action during movement – a component of lower extremity control often overlooked. Equal importance is placed on stretching, as well as control of movement.

Postural benefits

Improved alignment and integration of the hip, knee and ankle joints during bending; enhanced balance and postural control; increased strength and flexibility in bending.

Movement sequence

- Ankle mobility
- Calf stretch
- Hamstring stretch
- Hip stretch
- Hip flexor/quad stretch
- Deep squat stretch
- Single leg shoulder bridge
- Standing bow
- Wall squat pattern
- Single leg balance
- Squat progressions
- Lunge progressions

1. Ankle mobility

Refer to lesson 2, step 1.

2. Calf stretch

Refer to lesson 3, step 1.

3. Hamstring stretch

This stretch is practitioner-assisted, and makes use of contract-relax techniques.

The subject is supine, knees bent, feet standing, the practitioner holds one bent leg on their shoulder. The subject gently pulls their foot downwards onto the practitioner's shoulder to contract the hamstrings. This is done several times to establish a 'connection' with the muscle. Once this is done, the leg is straightened to a comfortable hamstring length and held by the practitioner. A mild stretch should be felt, but not a strain. The subject progressively pushes their leg downwards against the practitioner's shoulder, while simultaneously lifting the opposite foot off the surface (about 15cm). The subject should become aware of the hamstring and opposing hip flexors working together. This is held for five seconds. The subject then slowly releases the hamstring, while simultaneously putting the opposite leg back down. When the opposite foot touches the surface, the hamstring is relaxed.

At this point they should continue to push their opposite foot downwards into the surface – this will 'inhibit' the opposite hamstring so that the practitioner can take up the slack and stretch it further into a new position. From this new position, the cycle continues until the leg is completely straight.

Now a hand is placed in front of the thigh just above the knee. The subject then quickly pushes against this hand, i.e. contracting the quads to pull the leg upwards. Simultaneously, the practitioner begins to push the leg down slowly (like a drawbridge), overcoming the subject's force, until the leg is down on the surface. At this point, the leg is rolled gently side to side until loose. It is important that the subject does not over-exert the quads during this last phase, and that they progressively release the tension in the quads as they approach the surface. The cycle is repeated on the other side.

Fig 7.33 Hamstring stretch – hamstring contraction

Fig 7.34 Hamstring stretch – quad contraction

The arrow denotes the direction of push by the subject.

4. Hip stretch

Refer to lesson 4, step 2.

5. Hip flexor/quad stretch

Begin in a lunge position with your hands on the floor in front of you; place the foot and ankle of the leg to be stretched on a ball. As you slowly rise, you may begin to feel a stretch on the quads. If you cannot reach an upright position, you may keep you hands on the floor. Gently tilt your pelvis posteriorly (flatten back), until you reach a comfortable stretch, and hold for 30 seconds.

Fig 7.35 Hip flexor/quad stretch

6. Deep squat stretch

Stand in front of a fixed structure that you can grab hold of, such as a door frame. Squat as far as you can comfortably, allowing your back to round and your trunk to rest between your thighs. With your hands positioned through your legs, hold onto the fixed structure, so that you can relax further. In this position, gently rock back and forth from your heels to your toes for 30–60 seconds.

Fig 7.36 Deep squat stretch

7. Single leg shoulder bridge

Refer to lesson 4, step 8, for the basic movement.

Once this movement can be performed with good control, it can be performed with a single leg as follows: from the bridge position, allow your right hip to relax and push your left foot down into the floor – this will raise your left hip slightly higher. This is your start position: take a deep breath and as you exhale for a count of five, gently lower your left hip towards the floor by a few centimetres; simultaneously allow your right knee to drop out to the side by a few inches. Hold this position as you inhale for a count of five, and as you exhale, drop them both a little further towards the floor.

By repeating this movement with your breathing, you should aim to complete 5–6 repetitions before the left buttock and right knee reach the floor at the same time. Repeat the whole sequence three times, before switching legs.

Fig 7.37 Single leg shoulder bridge

The arrow denotes the leg dropping out to the floor.

8. Standing bow

Stand with your feet together and arms across chest. While maintaining an anterior pelvic tilt and good alignment, bend forwards from the hip, keeping your legs straight, until you feel a comfortable stretch in your hamstrings. Hold this for up to 60 seconds.

Fig 7.38 Standing bow

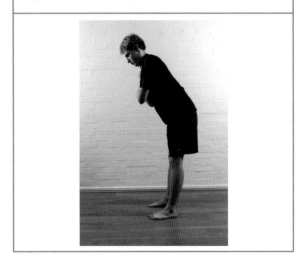

9. Wall squat pattern

Stand and place your hands on a wall, so that your elbows are almost straight. Position your feet about shoulder width apart, and ensure that your arms are at chest height. Begin to squat, as if about to sit on a chair, allowing your hands to slide down the wall with you. Return and repeat several times, slowly and methodically, making sure that you only squat to a comfortable level. Repeat the movement, keeping your hands fixed at chest height as you squat. Only descend as far as you feel comfortable.

Finally, perform the same movement, allowing one hand to slide up the wall and the other hand to slide down the wall. Perform several repetitions alternately on each side.

Fig 7.39 Wall pattern – hands fixed

10. Single leg balance

This movement is included here to stimulate the balance systems and somato-sensory awareness, in preparation for squatting and lunging progressions. It will also act as a useful active rest. Please refer to lesson 3, steps 5 and 6.

11. Squat progressions

The first movement begins to raise awareness of the feet during squatting, and stretches the extensor muscles. When standing, place your feet together, with the balls of your feet elevated on a 5cm block. Place a rolled up towel between your knees. Reach your arms up to the ceiling, and keep your legs straight, slowly reach down towards your toes, comfortably. Hold as you feel a stretch through the extensor muscles. Repeat 10 times.

Fig 7.40 Squat progression – toe touch with balls of feet raised

It is important that you reach only as far as is comfortable – squeezing the towel may help to relax muscles in your outer thigh as you bend. Repeat the movement again but with the heels raised on the block.

Fig 7.41 Squat progression – toe touch with heels raised

For the second movement, place your feet shoulder width apart, with your heels on the block. Place another block in front of you on the floor. As before, reach down with your legs straight until your palms are flat on the block. If your flexibility does not allow you to reach down with ease, then you may use a slightly larger block. Keeping your palms on the block, squat as far as you can go. Hold the bottom position for 10 seconds, as you breathe deeply. Notice that you can release tension in your legs and hips by pushing your hands into the block.

Fig 7.42 Squat progression – hands on block

After 10 seconds, reach your right hand up to the ceiling, turning your head towards your hand; push the left hand down into the floor to assist this. Hold this for 10 seconds breathing deeply. Repeat with the left arm. Bring both hands to the floor, and slowly squat up to standing. Repeat this up to six times each side.

Fig 7.43 Squat progression – single arm reach

In the final movement, begin by standing with your heels on the block, and reach your arms overhead. Slowly squat as deep as you can, taking about five seconds each way. Maintain a comfortable breathing pattern and repeat 10 times. Although this will feel challenging at first, practise will improve coordination.

Fig 7.44 Squat progression – double arm reach

12. Lunge progressions

Assume an in-line lunge position, where the heel of the lead leg is in line with the knee of the rear leg. Hold this for a few moments while the body stabilises. When ready, hold a wooden dowel across your shoulders. Now slowly begin to rotate your torso from left to right, taking about five seconds each way. Hold each rotation at the end range for five seconds. Repeat slowly, six times each side. During positions of instability, look for ways to re-align your posture for better balance, rather than engage the abdominal wall. Repeat on the other leg.

Fig 7.45 Lunge progression – in-line lunge balance

In the second movement, assume a split stance with the feet along the same line, about two foot lengths apart. The wooden dowel is held vertically along the back, with the hand of the front leg holding the bottom of the dowel, and the hand of the rear leg holding the top of the dowel. Once balance has stabilised, begin to lunge forwards onto the front leg, so that the back knee touches the ground. Hold this until your balance stabilises, then slowly return. Repeat up to 10 times on each leg.

Fig 7.46 Lunge progression – standing in-line lunge

8

LESSON PLANS FOR THE TRUNK AND SPINE

Lesson 7: Improving spinal mobility – flexion and extension

Theme(s)

A series of movements that bring awareness to the flexion/extension capabilities of the spine; higher integration is provided by progressive movements that demand increased spinal flexion and extension.

Postural benefits

Increased spinal mobility in the sagittal plane; improved strength and control of the trunk flexors and extensors.

Movement sequence

• Back and abdominal stretches
• Cat camel
• Relaxed pelvic tilt
• Supine knee push/pull
• Modified see-saw
• Floor push
• Seated flexion-extension

1. Back and abdominal stretches

To stretch the abdominal muscles, sit on a stability ball and slowly walk your legs out until you are lying on the ball with your knees bent. Allow as much of the spine to arch over the ball as is comfortable, and reach your arms overhead to increase the stretch through the abdominals. Hold for 30–60 seconds. If you suffer from dizziness of any sort, please ensure that your head is always positioned above your heart during this stretch.

Fig 8.1 Abdominal stretch

To stretch the back muscles, kneel behind the ball and slowly roll forwards onto your front, and relax in this position. Hold for 30-60 seconds breathing naturally. You may stretch different parts of your back, by rolling forwards or backwards, stretching the part of your back that is positioned over the apex of the ball.

Fig 8.2 Back stretch

2. Cat camel

Begin on all fours, with your hands shoulder width apart, and knees hip width apart. As you inhale for a count of five, push your hands into the floor, drop your head and flatten your back, allowing your spine to round upwards (flexion).

Upon exhalation, perform the opposite movement, as you gently lift the head, and arch the spine. Perform this movement slowly, allowing your awareness to focus on all vertebrae moving simultaneously, one way then the other. Repeat this movement for two minutes.

Fig 8.3 Cat camel

3. Relaxed pelvic tilt

Lie on your back with your knees bent, feet flat on the floor. Gently push your feet into the floor, as if trying to slide your feet away from you; Your pelvis begins to tilt posteriorly without the need for abdominal action. Practise this several times. Now from a neutral position, slowly begin to take the weight off your feet, without lifting them off the floor; your pelvis begins to tilt anteriorly without lower back effort. Practise this several times then combine the two movements. Use your hands to palpate the abdominal and back muscles noting how much more relaxed they feel as you tilt the pelvis. Spend up to two minutes on this movement.

4. Supine knee push/pull

Lie on your back and hold both your knees over the hip joints. Inhale as you gently push your knees forwards into your hands. Resist this movement, noticing the lower back muscles contracting, as the back arches. Your chin will also be pulled in slightly towards your chest. As you exhale, slowly release the knee push. Continue exhaling as you pull your knees towards your chest against the resistance of your hands.

You will feel your abdominal muscles contracting, your back flattening, and your chin turning upwards. Release once more, and continue the cycle for 10–15 repetitions.

Fig 8.4 Supine knee push/pull

5. Modified see-saw

The see-saw is an adaptation of the shoulder bridge movement.

Begin by lying on your back with your knees bent. Place your arms by your side. As you inhale, slowly lay your spine off the floor as you move towards the full bridge position. Exhale and slowly peel the spine back down and bring your chin towards your chest, just letting it go as far as comfortably possible. As you inhale again, peel the spine, and allow the head to lay back down. Repeat slowly 10 times.

Fig 8.5 Modified see-saw

6. Floor push

Lie on your front with your legs straight, head facing down. Position your hands just wider than shoulder width, in line with your chin. Take a breath in and, as you exhale, push yourself up as far as is comfortable, keeping your pelvis on the floor. Relax your buttocks and back muscles as you do this, focusing instead on the effort of your chest, shoulders and arms. Inhale as you return, and repeat slowly for 10 repetitions.

Fig 8.6 Floor push

7. Seated flexion–extension

In a relaxed sitting position, place your hands on your thighs. Slowly and comfortably raise your head and eyes as if looking towards the ceiling, then return and relax. Allow your back to arch slightly. Do this a few times noticing how much effort is involved, and how free the movement feels. How far can you see without straining the eyes or neck? This is your test movement.

Now, as you raise your head and arch your back slightly, simultaneously look down with your eyes. Go slowly and do this six times; the eye and head move in opposite directions. Go back to the test movement of moving the eyes and head the same way – can you see a little higher now? Does your spine move more easily?

Now, as you lower your head and round your spine slightly, allow your eyes to look upwards simultaneously. Return to the centre, and repeat slowly six times. Once again, return to the test movement. Can you see up a little further still?

Finish off by combining the above two movements – go slowly for six repetitions each way. Repeat the test movement at the end, noticing how much more freedom of movement you now have in extension and flexion.

When you have completed this sequence, notice how your spine feels. Perform a few flexion and extension movements of the spine, noticing any further differences in the way you feel. Stand up and walk around, paying attention to your alignment.

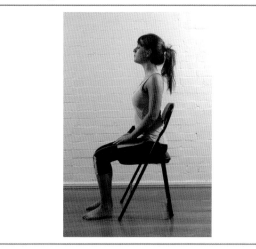

Fig 8.7 Seated flexion-extension – arching the back

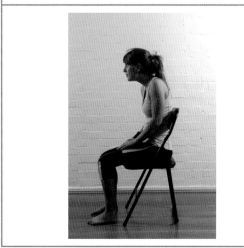

Fig 8.8 Seated flexion-extension – rounding the back

Lesson 8: Improving spinal mobility – side flexion

Theme(s)

A series of movements that bring awareness to the side flexion capabilities of the spine; further progressions are provided by movements that clarify the connection of the trunk to the hips and shoulders.

Postural benefits

Improved spinal mobility in the frontal plane.

Movement sequence

- Neck lateral stretch
- Side-lying ball stretch
- Supine slide
- Side-lying flexion
- Halo arms
- Supine reach

1. Neck lateral stretch

Sit on a chair in a relaxed upright position. Grab hold of one side of the chair seat and lean away until your shoulder is depressed. Using the other hand, reach over the top of your head and gently pull your head away from the fixed shoulder, until you feel a mild stretch in the side of your neck. From here, inhale as you push your head into your hand for a count of five. Slowly exhale and allow the stretch to increase slightly. Repeat twice and slowly return the head to the centre. Repeat on the other side.

Fig 8.9 Neck lateral stretch

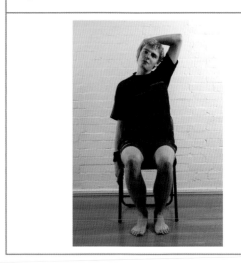

2. Side-lying ball stretch

From a seated position on a stability ball, roll down onto your back and then onto your side. Spread your legs wide enough to remain balanced; you may anchor your feet against a wall.

Reach over your head with your top arm, and grasp the wrist of your lower hand. Gently pull down the top arm until you feel a stretch through the muscles along your ribcage; you may wish to roll the upper body forwards or backwards a little to find a tight spot to stretch. Once you have found a tight spot, inhale and pull your upper body upwards against the pull of the lower arm. Hold for a count of five as you feel the muscles contract, then slowly exhale and release, allowing the trunk to stretch further. Repeat this cycle for any other tight spots, before swapping sides.

Fig 8.10 Side-lying ball stretch

3. Supine slide

Lie on your back with fingers interlaced behind your head. Your legs are bent with the soles of your feet touching. Begin by sliding your right knee up towards your right shoulder. As you do this, you will notice that your pelvis hikes a little and that your head wants to rotate to the right. Allow these movements to happen as you perform several repetitions slowly. Notice how your spine begins to side bend to the right.

Now rest your leg and focus on your right arm. Begin sliding your right elbow down the floor towards your right knee. Your head should move with your hand, and your spine should side-bend from the top half. Do this slowly several times.

The arrows denote the direction of limb movement.

Fig 8.11 Supine slide

Now combine the two movements, as you slide your knee up the floor and slide your elbow down the floor. Notice how much further you can side bend when the two movements are combined. Do this several times paying attention to the ease of movement and using as little effort as possible. Rest before repeating on the other side. Finally, combine the two movements from left to right slowly for 60 seconds. Relax and notice how your spine feels now compared to the beginning of the movement.

4. Side-lying flexion

The above movement can now be progressed to the side-lying position. Lie on your left side with your knees and hips bent to 90°. Your left arm is positioned straight overhead to support your head. The movement begins by lifting both feet up towards your head, as the thighs pivot around the knees. You will feel the right hip hike upwards and the right side of the waist shorten. Perform this movement several times noticing how much effort you are using. Rest the legs for a moment. Place your right hand on your forehead, so your elbow is pointing upwards, and begin to side bend the spine so that your elbow moves in an arc towards the right hip. Aim to keep your head in alignment with your shoulders.

The arrows denote the direction of limb movement.

By learning how to side bend from the thoracic spine, you will avoid having to side bend your neck. Perform this several times and rest. Now combine both movements – notice how the hiking of the hip assists in lifting the upper body, resulting in an easy side bend, without the need to bend the neck. Repeat several times using as little effort as possible. Rest and repeat on the other side.

Fig 8.12 Side-lying flexion

5. Halo arms

Lie on your back and bend your knees. Interlace your fingers above your head so your arms form a circle (not touching your head). Slowly slide your arms from left to right several times. Notice what is demanded of your upper back. Your head stays stationary. Try to make this movement smooth and notice how your spine, ribs, hips and breath are involved.

Imagine your head and arms are a unit and perform the same movement as above, sliding your head and arms together. Now let your head rest in the middle and just move your arms from side to side as in the first movement. Is it easier now? Rest, arms at sides – notice how contact with the floor is changing.

The arrows denote head and arm movement.

Now slowly slide the circle of your arms to the right, so your left upper arm is near your left ear, and stay there. From that position, slide your head from side to side on the floor, with your arms off-centre. Then change it, so instead of sliding your head, roll it comfortably and slowly from side to side. Repeat on the other side and then return to neutral, head in the middle, and slide the circle of your arms from side to side as at the beginning. Is it easier still?

The arrows denote head and arm movement.

Slide your head to the right while sliding the hoop of your arms to the left, then slide your head to the left and arms to the right. Repeat alternately many times, easily and comfortably. Rest and notice how your spine feels.

Fig 8.13 Halo arms – same side movement

Fig 8.14 Halo arms – opposite movement

6. Supine reach

Lie on your back with your legs straight and reach your right arm overhead on the floor. Now begin gently to reach up the floor with your right hand, so that you feel your arm slide several centimetres. Gently return under the recoil of the muscles and repeat slowly and deliberately several times. As you do this movement, relax your neck and notice which way your head wants to turn to make the movement feel freer. Add this to the reaching movement. You may now be aware of the rest of the spine moving in response to the neck.

The arrows show the sliding movement of the opposite leg and arm.

Now, relax your legs and notice what your pelvis has to do to reach the hand further. Allow the opposite side of the pelvis to hike up gently (as if sliding the opposite leg up the floor towards the shoulder) as you reach. Can you see how this allows you to go further into the reach? As this happens, you will also notice a subtle side bending of the spine away from the reach. Strange as this may seem, it's the opposing action of the hip that allows you to organise your body efficiently for reaching. You should feel a distinct stretching of the reaching side of the body. Continue this movement until it feels comfortable and more automatic. Repeat on the other side if necessary.

When you have completed this sequence, notice how the muscles of the side of your trunk feel. How does your ribcage feel? Is there any notable difference to your breathing? Move around, noticing any further differences in the way you feel.

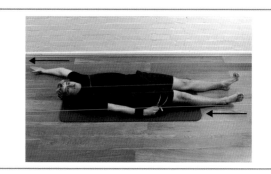

Fig 8.15 Supine reach

Lesson 9: Improving spinal mobility – rotation

Theme(s)

A series of movements that bring awareness to the rotational capabilities of the spine; higher integration is provided by progressive movements that incorporate differential rotation; also involves shifts in awareness to enhance movement. This simple sequence also understands that side flexion and rotation is coupled; as such this lesson should be performed alongside the previous one.

Postural benefits

Improved spinal mobility in the transverse plane; integration of the spine, shoulder and hip during rotation.

Movement sequence

- Neck rotation stretch
- Levator scapulae stretch
- Trunk rotation stretch
- Prone twist
- Supine twist
- Seated rotation

1. Neck rotation stretch

In a relaxed upright, seated position, rotate your head to one side (until you feel a mild stretch) and place the opposite hand on your cheek to hold this position. Inhale for five seconds as you look back towards the centre and push your cheek against your hand.

Resist this with your hand, then slowly exhale and look further behind you as you slowly increase the stretch on your neck. Use your hand to hold this new position and repeat two more times. Repeat on the other side.

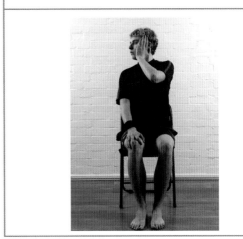

Fig 8.16 Neck rotation stretch

2. Levator scapulae stretch

Standing, reach one arm as far down between your shoulder blades as possible and look as far as you can to your opposite shoulder. Imagine you are trying to touch your chin to your opposite shoulder. In this position, the inside of your forearm should be making contact with the side of your head. Inhale for five seconds as you push the side of your head against your forearm – as if trying to bring your head back to the centre. Resist with your forearm then slowly exhale as you relax and look further towards the opposite shoulder.

Repeat this two more times, allowing your chin to drop a little further each time. Repeat on the other side.

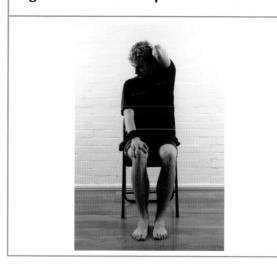

Fig 8.17 Levator scapulae stretch

3. Trunk rotation stretch

Lie on your back with your knees bent and your arms out to the side. Place your right hand on the outside of your right thigh and allow your legs to drop to that side until you feel a comfortable stretch in your lower back; maintain as much support of your legs as you can using your right hand. Inhale for a count of five as you slowly release the hold of your right hand – you will feel the trunk muscles contract. As you slowly exhale, allow your legs to roll further towards the floor, as your back stretches a little more. Allow your right hand to provide support for your legs once more, as your trunk muscles relax into a new position.

From here, repeat the contract–relax cycle until you can no longer improve your range of motion comfortably. Repeat on the other side.

Fig 8.18 Trunk rotation stretch

4. Prone twist

Lie on your front with your legs straight and forehead resting in your hands. Keeping your knees together, bend both legs so the soles of your feet are pointing upwards. Gently allow your feet to drop side to side, controlling the movement smoothly. Notice how your spine rotates as you do this, and pay attention to how far up your spine you feel movement. Do this for 60 seconds and rest the legs.

Now place your hands on the floor by your shoulders, and allow your elbows comfortably to point upwards. Begin to turn your head left to right as if looking under each arm towards your feet. Do this slowly and deliberately for 60 seconds. Notice how your neck and shoulders rotate. Follow this movement as far down your spine as you can. Try to move with as little effort as possible.

Finally, combine the two movements – as you drop your feet to one side, turn your head to look through the arm on the opposite side. Continue this movement fluidly and slowly from left to right for 60 seconds. Notice how the counter-rotation of the upper and lower spine reinforce one another to make the whole movement feel natural and coordinated. Use as little effort as possible.

Fig 8.19 Prone twist – head movement

Fig 8.20 Prone twist – combined movement

5. Supine twist

Lie on your back with your legs bent. Allow your head to rotate gently from left to right, pivoting around the point that is in contact with the floor. Allow the weight of your head and gravity to guide the movement, with least effort.

Now hold opposite elbows with your hands and begin to guide each elbow towards the floor as your shoulders begin to roll left and right. Focus on keeping the movement smooth and coordinated. You may also notice how your head rolls all by itself, as the shoulders move, allowing for even freer movement of your neck.

Now allow your eyes to move in each direction, just before your shoulders move. Notice how your eyes set a direction for the whole body.

Lastly, allow your knees to drop gently side to side in opposition to your arms and head.

Notice how your feet roll; how your knees move; and how your pelvis rocks and tilts.

Perform the entire movement for several minutes with good awareness, and the least effort necessary, before resting.

Fig 8.21 Supine twist

6. Seated rotation

Begin in an upright, seated position towards the front part of a chair. Place your hands on your thighs. Slowly turn your upper body as if comfortably looking over your right shoulder – do the movement without strain. Notice how far you go. Now keeping your body still and your eyes facing forwards, turn your head only to your right and back again. Do this slowly six times – only your head moves, your eyes remain looking forward. Rest a moment. Now allow your shoulders and spine to rotate to the right while keeping your head and eyes facing forwards. Do this slowly six times.

The arrows denote the head turning back and the shoulder turning forward.

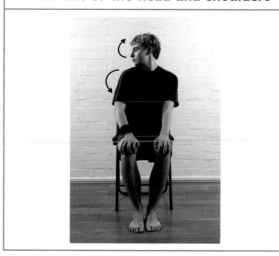

Fig 8.22 Seated rotation – opposing movement of the head and shoulders

Re-test using the original movement – notice how turning to the right is becoming easier and more comfortable.

Now with your feet remaining flat on the floor, move your left knee forwards slightly; relax your left foot and leg as much as possible. Notice how this movement begins to rotate your pelvis, back, shoulders and head to the right. Notice how you also grow a little taller as you rotate. Do this several times until it feels co-ordinated.

The arrow shows the leg pushing forward.

Re-test using the original movement – notice how you can turn a little further still to the right.

Now, combine all the above movements; as you push your left knee forwards, turn your shoulders, head and eyes to the right. Notice how easy the movement is now compared to the beginning. Do this several times, until the movement feels natural without strain. Imagining that you are actually looking at something worth looking at will further enhance your awareness of this movement.

Finally, rest and compare how one side of your body feels in relation to the other. How does the neck feel? How do the shoulders feel? How does the spine feel? Repeat this sequence on the left side.

Fig 8.23 Seated rotation – movement of the knee

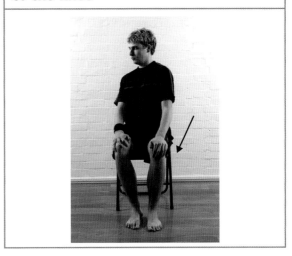

LESSON PLANS FOR THE NECK AND SHOULDERS

9

Lesson 10: Freeing up the shoulders

Theme(s)

An exploration of movements that clarify the relationships between the shoulder joints, scapulae, thoracic spine and hip joints. By establishing the many connections between these structures, the shoulder is able to mobilise with less effort, therefore freeing up any restrictions that may be present. Because the shoulder has a large number of muscle attachments, the first half of this sequence uses a variety of stretching and release techniques in order to enhance movement awareness.

Postural benefits

Increased freedom of shoulder movement and reduced muscular tension; improved scapula control; better alignment of the shoulder joint and girdle; improved head alignment.

Movement sequence

- Latissimus stretch
- Chest stretch
- Rhomboid stretch
- Levator scapulae stretch
- Side-lying teres stretch
- Release techniques (assisted)
- Side-lying rotations
- Side-lying reach
- Prone overhead slide
- Supine twist
- Side roll

1. Latissimus stretch

Kneel in front of a ball and bend your arms so you can rest the elbows on the ball. Your elbows should be positioned as close together as possible. Rest your head in the crook of your elbows and slowly sit back towards your heels as far as is comfortable. When you feel a stretch down the sides of your back, hold this position while you slowly tilt the pelvis posteriorly (flatten lower back). Hold this for 30–60 seconds.

Fig 9.1 Latissimus stretch

2. Chest stretch

Kneel on all fours and place one arm on the ball. Keep your shoulders parallel to the ground and slowly drop your body towards the floor until you feel a mild stretch in the chest. Inhale for a count of five as you push your arm down into the top of the ball. You will feel your chest muscles contract. Exhale slowly as you drop your body further into the stretch. Repeat once more from this new position, and hold the final position for 10 seconds before returning slowly. Repeat on the other side.

Fig 9.2 Chest stretch

3. Rhomboid stretch

Kneel in front of a ball and place your elbow on the ball. Allow your body weight to come forwards and rest through your upper arm, so that you begin to feel a stretch through the back of your shoulder. Inhale for a count of five as you push the elbow into the ball (as if trying to pull your shoulder blade back). Slowly exhale and stretch further as you allow the shoulder blade to move away from the spine. Repeat the cycle once more, before swapping sides.

Fig 9.3 Rhomboid stretch

4. Levator scapulae stretch

Refer to lesson 9, step 2.

5. Side-lying teres stretch

This stretch differentiates the latissimus from the teres muscles. Lie on your side with your knees bent and lower arm straight, to support your head. Place a rolled up towel underneath your shoulder at the level of the scapula, so that the scapula is pushed back against the spine. Relax and hold this position for 30–60 seconds. Repeat on the other side.

Fig 9.4 Side-lying teres stretch

6. Release techniques (assisted)

The subject lies supine with their knees bent comfortably. The practitioner picks up one arm and begins to move the shoulder joint through flexion, extension, abduction and adduction, and rotation. The movements should be performed slowly, so that the subject can fully relax the arm and shoulder. The practitioner should feel for restrictions as well as freedom of movement. These small movements are made for up to two minutes, or until the movements feel freer. Similar movements can also be made in a side-lying position.

Fig 9.5 Supine release

With the subject lying prone, their head is turned to the side that is being released. The same side elbow is flexed and their shoulder is moved gently into internal rotation, so that the hand can be positioned in the lower back. It is held there by the practitioner's knee, underneath the elbow. The practitioner's hand(s) can now be used to gently mobilise the scapula.

The subject will soon be able to release any habitual contractions in this area as they relax.

Fig 9.6 Prone release

7. Side-lying rotations

Lie on your left side, so that your knees are bent 90°; you can place a small foam roller between the knees to relax the hip muscles. Rest your head on your upper arm. Your right elbow should be bent, and your right palm resting on the floor in front of your chest.

Place your right hand in a position where your elbow is pointing towards the ceiling, as if balanced in space. Begin slowly to shrug your right shoulder up towards your right ear and back again. Do this several times. Now begin to move your elbow towards your head instead. Notice that your shoulder shrugs upwards, with very little effort from your neck muscles. Notice how this movement also pulls the ribcage upwards – which in turn begins to tug gently at the waist muscles. Continue this movement for 30–60 seconds.

In the same way, begin to explore a movement of pushing your shoulder away from your body and back behind your body (protraction and retraction) noticing how your head and spine also rotate. Finally, you can combine these movements to make circles with your shoulder joint. Do it slowly and deliberately, paying attention to how your head and spine move with your shoulder. Relax for a moment, noticing how the shoulder feels.

The arrow denotes the rotation of the shoulder in a circular fashion.

Perform these same movements with your hip, first moving it up and down, then forwards and backwards, and finally in circles. See if you can perform the movements with the least amount of muscle contraction. Notice what your shoulder is doing as you move your hip.

The arrow denotes the rotation of the hip in a circular fashion.

Fig 9.7 Side-lying rotations – shoulder rotations

Fig 9.8 Side-lying rotations – hip rotations

Spend a few minutes exploring the different ways in which you can combine movements of the shoulder and hip. Explore movements in the same direction, as well as those in opposing directions. Notice how movements of the shoulder require integration of the hip. Aim to move with as little effort as possible, keeping your movements small yet fluid. Turn onto your back and notice how the right side of your torso feels compared to the left. Repeat on the other side.

8. Side-lying reach

Lie on your left side as before and place your right arm straight out in front of you on the floor, palm down. Allow your shoulder and neck to relax. Begin by turning your eyes to look down to the floor, while simultaneously relaxing your right arm. As you do this, you will begin to notice your head and shoulder rotating as the palm slides away from your body. Notice how easily this happens then return using as little effort as possible. Repeat this several times, paying attention to how the thoracic spine rotates and helps the shoulder to reach further forwards.

The arrow denotes the sliding of the arm away from the body.

Now return to the start position and perform the same movement as you rotate the shoulder backwards. Notice how the eyes and head turn towards the ceiling and the palm slides nearer to you. Lastly, combine both movements, integrating your eyes, head, shoulder and spine. Explore this movement forwards and backwards for 1–2 minutes, before swapping sides.

9. Prone overhead slide

Lie on your front with your head resting in your left hand, facing right. Place your right arm beside your head, with your forearm resting on a foam roller. Keeping your arm and shoulder as relaxed as possible, begin to reach up the floor with your hand, allowing your arm to straighten as it rolls. Reach as far as is comfortable before slowly returning. Allow the roller to bear the weight of your arm completely, and repeat several times. Repeat on the other side.

The arrow denotes the sliding of the forearm on the roller.

Fig 9.9 Side-lying reach

Fig 9.10 Prone overhead slide

10. Supine twist

Refer to lesson 9, step 5.

11. Side roll

Lie on your back with your legs straight and your arms by your side. Bend your left leg and begin a movement of pushing your left foot into the floor. Notice how your left buttock begins to lift and your body rotates to the right. Push a little more and notice that rotating your head helps this movement. Now begin to slide your right arm up the floor towards a horizontal position, as you rotate with as little effort as possible.

Fig 9.11 Side roll – start position

Now, allow your eyes to turn your head towards a point just above your head; reach for this point with your left arm. Notice how this pulls your whole body in rotation effortlessly, so that you roll onto your right side. Repeat this movement slowly for two minutes each side, noticing how your hip, pelvis, shoulder and spine work together to produce coordinated movement.

Fig 9.12 Side roll – end position

Lesson 11: Freeing up the neck

Theme(s)

An exploration of movements that clarify the relationships between the eyes, neck, shoulders and trunk. A number of possible movements of the neck are explored, which are then progressively integrated with actions of the eyes, shoulders and trunk, allowing the neck to release itself from unnecessary effort.

Postural benefits

Increased range of movement; reduced muscular tension; improved head carriage.

Movement sequence

- Head nod stretch
- Neck rotation stretch
- Neck lateral stretch
- Neck release (assisted)
- Supine horizontal arm reach
- Halo arms
- Head hold with abdominal integration
- Side roll

---

1. Head nod stretch

In an upright, seated position, nod your head down towards your chest. Place one hand on your chin and one hand on the back of your head. Gently stretch the back of your neck by pulling the back of your head upwards, and pushing your chin inwards. Inhale for a count of five, push your chin against your hand as you resist with the hand. Slowly exhale for five, as you release the hand and gently pull the top of your head upwards to stretch further. Repeat this cycle two more times.

The arrow denotes the direction of stretch.

2. Neck rotation stretch

Refer to lesson 9, step 1.

3. Neck lateral stretch

Refer to lesson 8, step 1.

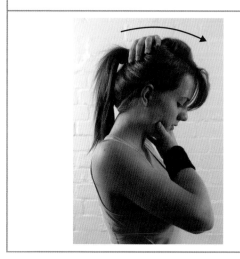

Fig 9.13 Head nod stretch

4. Neck release (assisted)

The subject lies supine and the practitioner cradles their head using both hands, so that it is about 1–2 inches above the surface of the couch. It is important that the subject feels confident enough to release all efforts of holding the head using the neck muscles. This position is held for 60–90 seconds.

5. Supine horizontal arm reach

Begin in a supine position with your legs bent and arms out to the side. Begin to reach your arms further outwards, as if trying to reach for something that is just beyond your reach. Think about reaching from the chest and shoulders, with a small movement. As you hold this position, you will feel a distinct pulling apart of your shoulder blades. Notice your head position now. You may feel as though the back of your neck is closer to the floor and your chin has moved towards the chest.

6. Halo arms

Refer to lesson 8, step 5.

7. Head hold with abdominal integration

Lie on your back with your legs bent, arms by your side. Gently nod your chin towards your chest and slowly lift your head off the floor slightly. Notice how your neck flexors contract.

Slowly return and do this movement several times without straining too much. During this movement, you may begin to notice your neck extensors relaxing as they are reciprocally inhibited by the action of the neck flexors.

Fig 9.14 Neck release

Slowly release your arms from this position and repeat the movement slowly until your neck muscles begin to relax automatically with the movement.

Practise this for 1–2 minutes then rest and notice how the neck feels.

Fig 9.15 Head hold with abdominal integration – chin tuck

Now flatten your back into the floor and lift your knees off the floor over the belly. Hold this for a few moments, noticing the contraction in the abdominal muscles. Release and repeat several times.

Finally, combine the two movements. Tuck your chin in and lift your head up (no more than 5–8cm off the floor), while simultaneously bringing your knees inwards to flatten your back. Notice how the neck flexion facilitates the trunk flexion and vice versa. Use your eyes to facilitate the neck movement by looking down towards your belly. Hold this position for up to 30 seconds, as you breathe freely and relax all unnecessary efforts. Slowly return the legs and head to the floor.

The moment the head touches the floor you may notice a relaxation of the neck extensors. Repeat the movement two more times. Lie back and notice how your neck muscles feel.

8. Side roll

Refer to lesson 10, step 11.

Upon completion of this lesson, take the time to stand up and notice how your neck feels, especially in relation to your eyes, shoulders and spine. For those using this lesson to re-align the head, you will undoubtedly notice that the head is now sitting further back on your shoulders with little effort. As your nervous system processes this information, notice the alignment of your shoulders on your ribcage, and the ribcage on the pelvis. If you are using this lesson simply to improve neck mobility, you will notice an increased sense of awareness of head carriage, which you can continue to build on.

Fig 9.16 Head hold with abdominal integration – knee tuck

Fig 9.17 Head hold with abdominal integration – combination movement

Lesson 12: Enhancing control of the shoulder muscles

Theme(s)

Explore how the muscles of the scapula, shoulder joint and cervico-thoracic spine can interact to produce optimal alignment and co-ordinated movement. This lesson begins to focus attention on a number of important force-couple relationships in the shoulder. You will also be more aware of the process and quality of muscular contractions which will improve the control of key muscles responsible for optimal alignment.

Postural benefits

Improved strength and control of the scapulo-spinal and scapulo-humeral muscles; increased awareness of the shoulder muscles and how they are designed to move; increased mobility of the scapulae.

Movement sequence

- Release techniques (assisted)
- Neck stretches
- Latissimus, teres major, rhomboid and chest stretch
- Prone scapula pulls
- Four-point rocking
- Serratus pulls
- Overhead wall slide
- Seated push up

1. Release techniques (assisted)

Refer to lesson 10, step 6.

2. Neck stretches

Refer to lesson 11, steps 1, 2 and 3.

3. Latissimus, teres major, rhomboid and chest stretch

Refer to lesson 10, steps 1, 2, 3 and 5.

4. Prone scapula pulls

Lie on your front with forearms resting on a foam roller, about shoulder width apart. Allow your head to rest down. As you inhale, begin slowly to reach your arms overhead, under control. Exhale as you slowly release and continue to exhale as you pull your arms towards you.

The arrows denote reaching forward with the hands.

Fig 9.18 Prone scapula pulls – elevation

Continue elevating and depressing the scapulae in this way 10 times, noticing the muscle contractions taking place across your shoulders and back, but do not use excessive effort.

The arrows denote pulling back the arms and lifting the chest upwards.

Allow the head to lift up gently as you pull your arms down, and return as you reach up. Notice how the head movement facilitates the shoulder movement.

5. Four-point rocking

Refer to lesson 17, step 4.

6. Serratus pulls

Kneel on all fours with your wrists positioned under your shoulders. Ensure that there is even weight between your knees and hands, and your spine is in a neutral position. As you inhale for a count of five, push your hands into the floor, keeping your arms straight. Feel your chest rising and your upper back moving towards the ceiling. Notice how your shoulder blades are pulled apart, and feel a tightness underneath the pectoral muscles. Do not allow your shoulders to elevate. Hold this position for a moment, then slowly release as you exhale, allowing your chest to fall under its own weight – keep your arms straight and vertically aligned throughout. Repeat 10 times slowly.

The arrows denote pushing the arms into the floor and pushing the upper back up.

Fig 9.19 Prone scapula pulls – depression

Fig 9.20 Serratus pulls

7. Overhead wall slide

Stand facing a wall with your feet about 15cm away. Lean forwards slightly and rest your forearms flat on the wall. Your hands should be positioned at chin height. Your arms will form a 'W' shape with your head. Keeping your hands in contact with the wall, begin to slide them up as you progressively shrug your shoulders. Use the friction of the surface as resistance for the movement, so you feel the muscular effort in your shoulders and upper back. Inhale slowly, reaching as far as you can. Hold this position feeling the contraction of the upper trapezius and deltoid muscles.

As you exhale, allow your shoulders to fall slowly under their own weight – keep your arms straight as you do this. Feel the same muscles relaxing. Notice how your chest elevates slightly on the return. Inhale and shrug up from this position, and continue the movement for 10 repetitions, using as much resistance as you can while maintaining good control.

8. Seated push up

Sit on a stability ball with good posture. Place your palms on the ball by your side in line with your hip joints. Relax your shoulders and begin a movement of slightly depressing the scapulae; as you continue this movement, push your hands down into the ball and straighten your arms as much as you can. Keep the scapulae depressed as you do this, visualising the scapulae moving downwards and slightly inwards.

You will feel muscles in the mid-lower back contracting (lower trapezius). Exhale during this part of the movement. Inhale deeply on the return and repeat 10 times.

Following this lesson, notice how the muscles of your shoulder and scapulae feel. Are the scapulae positioned differently, and how does this affect the position of your chest and head?

Fig 9.21 Overhead wall slide

The arrows denote sliding the arms up the wall.

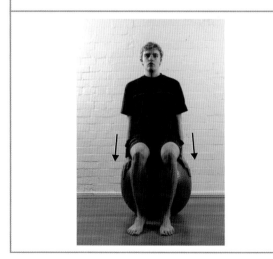

Fig 9.22 Seated push up – end position

The arrows denote pushing the arms down into the ball.

LESSON PLANS FOR BALANCE AND ALIGNMENT

10

Lesson 13: Better balance

Theme(s)

Exploring single leg balance from the ground up. This sequence of movements focuses on improving single leg balance without the need for excessive use of the abdominal wall. Instead it draws attention to joint alignment and the maintenance of this alignment by the tensional characteristics of soft tissue structures. As the nervous system begins to learn how to release unnecessary effort, a new set of control mechanisms will evolve that can effectively be transferred to other areas of postural control.

Postural benefits

Enhanced sense of balance, stability and security; improved alignment governed by tension, rather than compression; reduction in unnecessary use of the abdominal wall for stability.

Movement sequence

- Establishing a point of reference
- In-line lunge balance
- Three-point foot stance
- Single leg lengthening
- Single leg calf raises
- Single leg knee bends
- Re-test

1. Establishing a point of reference

This is an important first step which will be re-tested at the end of the sequence to obtain feedback. Stand on one leg and balance for as long as you can. Make a note of how this movement feels in terms of overall stability, effort used and length of balance. Do this no more than three times, and do not repeat on the other leg; you will need to work through the rest of the sequence using the same leg. The whole sequence can be repeated on the other leg afterwards.

2. In-line lunge balance

The aim of this movement is to begin a process of recalibration so that your body can learn to mobilise the trunk and balance simultaneously. Refer to lesson 7, step 12 (first movement) for a description of the in-line lunge balance.

3. Three-point foot stance

Stand on your test leg, using support if necessary. Allow your body to relax but remain upright with good posture. Focus your attention to your support foot. Imagine three points on the sole of the foot: one in the heel, one at the base of the little toe and one at the base of the big toe. These points form a triangle. Think about applying equal pressure through all three points, so that your body weight feels balanced over the centre of the triangle. Use as little support as you can. If you are unnecessarily contracting the foot muscles to hold this position, imagine the three points moving further apart, so that the triangle gets slightly bigger. Practise this for 1–2 minutes.

During this process, be aware of the sensations in your foot – you may notice your toes spreading as you move the three points apart – this is a more natural foot position that is conducive to good balance.

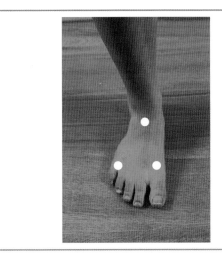

Fig 10.1 Three-point foot stance

4. Single leg lengthening

Standing as before, visualise the three points and hold this position. Bring your attention to the centre of the triangle and imagine pushing down through this point. As you do this, notice how your entire leg begins to lengthen upwards and the opposite side of your pelvis hikes a little. Hold this position for 10 seconds using as little effort as possible, focusing on increasing the apparent length of your leg. Rest and repeat five more times.

Progress this movement further by allowing your hip to hike up, and slowly allowing it to fall below the level of the support hip. You can then hike it back up again, and continue the cycle for 10 slow repetitions. During this progression you will become aware of the gluteus medius muscle shortening and lengthening in the support leg.

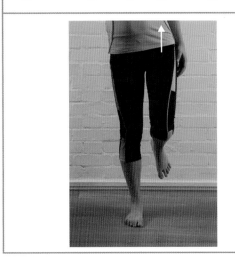

Fig 10.2 Single leg lengthening

5. Single leg calf raises

Stand on the same leg as before and lengthen through the leg. Focus on the two points of the triangle at the ball of your foot and an imaginary point in the middle of your chest. Simultaneously, push the points on the foot into the floor and lift upwards through the point in your chest, as the heels rise and your body weight moves onto the balls of your feet. Allow your heels to lift as far as feels comfortable – you will feel your calves contract. Find a sense of balance in this position, using the support accordingly. Slowly return and repeat 10 times.

Aim to maintain the awareness of length through the leg as you do this movement, returning back to the three-point stance each time.

6. Single leg knee bends

Stand on the same leg and lengthen through as before. Slowly perform 10 knee bends. As you do this, only bend your knee as far as it remains over the foot and ensure that your knee remains in line with the second toe throughout. Aim for equal balance over the three points on the sole of the foot and keep your leg lengthened by pushing the floor away from you with each push. Allow your gaze to drop slightly as you bend down and return to horizontal as you rise.

Perform this movement without support if you feel comfortable.

7. Re-test

Repeat step 1. Notice how your balance feels now. How much effort are you using? Can you hold this position for longer? Take your time to notice these changes.

As you repeat this lesson, aim to perform the movements without using additional support. Initially, you may experience over-use of the abdominal muscles to keep you balanced; to overcome this, you may return to step 2 as often as you like, until you begin to feel comfortable in balancing. As this recalibration takes place, you will undoubtedly increase your control of balance using less effort. Look for ways to practise the single leg balance in everyday life, so that changes can also take place more unconsciously.

Fig 10.3 Single leg calf raise

Fig 10.4 Single leg knee bends

Lesson 14: Re-educating the oculo-vestibular system

Theme(s)

Exploring some important movements that challenge the visual and vestibular systems with the aim of improving and maintaining postural control. While the movements are designed to work in sequence, they can be done individually.

Postural benefits

Enhanced static, dynamic and reactive stabilisation; improved alignment.

Movement sequence

- Eye tracking
- Gaze stabilisation
- Peripheral awareness
- Ball tosses
- Heel and toe walking
- Single leg balance progressions
- Balance board progressions

1. Eye tracking

In a seated position with good posture, keep your head looking forward. Begin slowly tracking your eyes left to right, taking about two seconds each way. Do this 10–20 times, staying within a range of motion that feels unstrained. Rest by closing your eyes for a few moments.

Now perform the same movement with your eyes moving up and down.

As your control improves, progressively increase the speed of movement to one second each way, and then half a second each way. Other variations include diagonal and circular movements. You may also progress to doing these standing.

If you experience any dizziness, eye strain or headaches during these movements, stop immediately and rest.

2. Gaze stabilisation

These movements require your eyes remain fixed forwards as your head moves. Use the same movements and progressions as in step 1: side to side, up and down, diagonal and circular, increasing speed as you improve control.

3. Peripheral awareness

Stand with one arm straight out in front, thumb pointing up. Keep your eyes and head facing forwards. Close one eye and slowly sweep your arm out to the side until you lose sight of your thumb in your peripheral vision. Slowly return until you catch sight of the thumb once more. Repeat this movement slowly six times. Repeat with the other eye closed and then repeat with both eyes open.

Perform the movement again, but this time keep both your eyes open and allow your head to rotate away from the moving arm. This will begin to calibrate your peripheral vision during head and target movement. Do the movements slowly and deliberately without strain.

You may also wish to explore other arm movements such as up and down, or circles.

Fig 10.5 Peripheral awareness

4. Ball tosses

Standing, hold a small ball in one hand and throw it back and forth from one hand to the other. Begin with your hands about 12 inches apart and positioned at chest height. Do this for about one minute; then do the same at head height and hip height.

Now go through the same variations, but this time, keep your gaze fixed forwards. How does this change your ability to throw and catch? As you begin to gain better control, move your hands further apart and/or increase the speed of throwing. Other variations include one eye closed, throwing the ball against a wall, and juggling.

Fig 10.6 Ball tosses

5. Heel and toe walking

Refer to lesson 3, step 4.

6. Single leg balance progressions

Refer to lesson 3, step 5; and lesson 13.

7. Balance board progressions

Refer to lesson 3, step 6.

Following this lesson, stand up and move around, noticing how your sense of balance feels. How do your eyes feel? Look around you and into the distance to gauge any differences in your vision.

Lesson 15: Organising the body from lying to standing

Theme(s)

Sequenced rotational movements are used to explore the process of moving from a lying to standing position, and back; this lesson offers a wonderful example of how the body can organise itself efficiently as it moves from one position to another. Good use is made of rotational patterns, and integrated use of the eyes and head to set a direction for movement.

This sequence consists of five distinct movements: supine to prone roll, prone lying to sit, sit to kneel, kneel to lunge, lunge to stand. While these movements are described individually, they will eventually be combined together to form one fluid sequence. Feel free to explore each one individually before performing the entire sequence.

Postural benefits

Improved sense of balance, coordination and strength.

Movement sequence

- Supine to prone roll
- Prone lying to sit
- Sit to kneel
- Kneel to lunge
- Lunge to stand

1. Supine to prone roll

Lie on your back with your legs straight and arms by your side. Bend your left leg and begin a movement of pushing your left foot into the floor. Notice how your left buttock begins to lift and your body starts to rotate to the right.

Fig 10.7 Supine to prone roll – start position

As you continue to push a little more, you may begin to notice that your head also wants to rotate – allow it to do so, noticing how it assists in the movement. As you continue, begin to think about what else has to move for you to roll onto your right side. Begin to slide your right arm up the floor as you push the left foot. Does this help the movement? Use as little effort as possible.

Now allow your eyes to look at a point above your head and reach your left arm towards this point. Notice how your whole body begins to turn effortlessly onto the right side and then onto the front. As the movement becomes familiar, try to find ways of performing it with as little effort as possible. Repeat this movement back and forth slowly for two minutes each side noticing how the hip, pelvis, shoulder and spine work together to produce coordinated movement. Now, remain on your front.

2. Prone lying to sitting

In this position, with your arms out straight, imagine you are about to push yourself up from the floor. Your aim is to move into a side-sitting position using as little effort as possible. To help do this, you will be rotating in the opposite direction from which you have just come; as you have rotated right, onto your front, you will now rotate left to a seated position.

Fig 10.8 Supine to prone roll – mid movement

Fig 10.9 Prone lying to sitting – start position

As you think about pushing up, notice where your hands position themselves – probably somewhere at chest level to offer maximum leverage for pushing. Allow your legs to bend at the knees – this will reduce the effort required to rotate. Now, begin to turn your eyes to the left, as if looking over your left shoulder – notice how your head follows and your body begins to turn. Your legs will begin to counter-rotate to the right and as you think about coming to an upright position, your arms will naturally straighten as you push into the floor. As you push, allow your left leg to move further back from the right, so that your legs move apart.

Use your eyes to pull you upwards, and you will now find yourself in an upright side-sitting position, with hands on the floor in front of you for support. Your right knee will be in front of you, and the left one behind you. Notice how you reached this position using your eyes and rotating the opposite way to the last movement. Reverse the movement and practise back and forth a few times slowly until it feels effortless. Integrate this movement with the previous one, come back to the side-sitting position and rest.

3. Sitting to kneeling

From the sitting position with your right knee forward, you shall now be rotating to the right towards a four-point kneeling position.

Fig 10.10 Prone lying to sitting – mid movement

Fig 10.11 Sitting to kneeling – start movement

Think about the end position of being on all fours – notice how your right hand and right knee are close to their end positions. How do you have to organise your left side? As before, allow your eyes to help you – begin to turn them towards the right and allow your head and body to follow you. As you do, your left hand reaches for the floor and your body pivots on your right knee, allowing your left knee to lift over your right knee to position on the floor.

You are now on all fours. Practise back and forth slowly and integrate it with the previous movements. Come back to all fours and rest.

Fig 10.12 Sitting to kneeling – mid movement

4. Kneeling to lunge

Having just rotated to the right, now rotate to the left to come to a kneeling lunge position.

Moving into this position requires a counter-rotation of the hips – as you pivot on the ball of the left foot, lift the left knee and turn it to the left; simultaneously pivot on the right knee and turn the right foot to the right.

Fig 10.13 Kneeling to lunge – start position

Turn your eyes 90° left to provide the momentum for this movement and use your hands to push up from the floor as your body rotates left into a lunge position. Practise back and forth slowly, and integrate with all previous movements. Return to the lunge position and rest.

Fig 10.14 Kneeling to lunge – mid position

5. Lunge to stand

This time the objective is to rotate to the right to come to standing. Notice where your feet are positioned – this is where they will remain throughout this movement. Think about a final standing position with your feet this distance apart. Now begin to look to where you are going – this will not only involve turning your eyes to the right but also upwards, a movement that will also turn and lift your head. As this happens, think about moving your centre of gravity in the same direction – this will mean transferring your body weight away from the front leg and towards the back leg.

Begin to focus on rotating the trunk to the right as it lifts and you will soon find that your body effortlessly positions itself equally over both feet. This is helped by rotating around the left heel and the ball of the right foot. You will now be facing at least 90° to the right of your last position. Bring the feet together slowly and relax to stand. Practise this back and forth until it feel effortless and smooth.

You may now practise the whole sequence from lying to standing and back, on both sides. Do this several times and notice how you feel afterwards.

The more you practise this sequence, the less you will think about the individual movements, and the more you will become aware of how important it is to organise the body during movement. There is no right or wrong way to do this movement – only a way that works for you. You will begin to notice how easy it is to move from one position to another simply by using your eyes to motivate the rest of your body. This mechanism is part of a deeply-rooted pattern that motivated us to reach forwards, and to crawl; to reach upwards, and subsequently learn to stand; and to reach forwards once more, and subsequently learn to walk. As you progress you will undoubtedly discover subtle variations within this sequence that will allow you to move with less effort.

Fig 10.15 Lunge to stand – start position

Fig 10.16 Lunge to stand – mid position

Fig 10.17 Final standing position

Lesson 16: Postural strings

Theme(s)

Moving towards optimal alignment using dynamic imagery. The visualisations in this lesson (and the movements they represent) form part of an on-going strategy that can be practised all the time. As such, it is an important beginner's strategy, as well as a long-term one. It should be periodically re-tested to evaluate whether lasting changes are being made.

Postural benefits

Enhanced awareness of optimal joint alignment with respect to gravity; reduced muscular tension and effort in static and dynamic postures.

Movement sequence

Just one movement.

Postural strings

Stand and imagine there are three pieces of string attached to your body: one at the top of the sacrum, one at the base of the neck and one on the forehead.

Now imagine that each string is being gently tugged upwards and away from your body. As you pull the string at the sacrum, you will notice a lengthening sensation as your pelvis tilts slightly anteriorly. The aim is to allow this to happen rather than cause it to happen. There should be no strain as you do this. As you pull on the string at the back of your neck, the back of your head moves up and your chin moves down and slightly inwards towards your throat. Do not pull in to form a double chin. Finally, as you pull on the string on the forehead, it will counterbalance the downward movement of your chin so that your head remains lightly balanced on your shoulders and your neck feels relaxed and elongated. Imagine your eye orbits are positioned horizontally. When your head is aligned in this way, you will also notice a natural lifting of your chest and your ribs will expand, creating a narrower waist. Notice how this happens with little muscular effort.

Fig 10.18 Postural strings – side view

The arrows denote the direction of pull of the imaginary strings.

When done with the minimal of effort, you may also experience lightness through your feet, as your weight distribution adjusts to a more central position over your feet.

Some more strings may be added to your body: imagine a string on each shoulder which when pulled laterally, help to widen your shoulders and expand your chest slightly. You will notice that this provides more lift to the chest and relaxes your neck muscles so that your head can sit back with ease. Imagine that your clavicles are sliding horizontally outwards and you will feel your chest and ribs open up more. Imagine a string with a small weight attached to the pubic bone. As this weight hangs between your legs, it pulls your pelvis/spine downwards, while the other strings pull you up. The overall effect may make you feel grounded or centred.

Lastly, imagine that you are standing in a swimming pool or in zero gravity. Notice how there would be less compressive force coming through your hip joints and how your legs would hang easily from the pelvis – almost like a puppet's legs hang freely. As you do this, you may notice that the leg muscles soften, your pelvis tilts slightly anteriorly and you sway a little. Notice how much more mobility you feel through your pelvis and low back.

Hold this posture for a few moments, taking time to get as much feedback as possible. Allow your body to settle back into the previous posture. Notice the differences immediately – you now have two points of reference – a posture that is optimal and one that is less than optimal. As you practise aligning yourself in this way, your nervous system will soon learn to adapt to a position that is less stressful – a position that will become the preferred way of aligning your body. However, it is important that you practise this with good awareness and feedback to facilitate learning. Once the above posture is achieved, learn to integrate it into all daily activities.

It is important for the practitioner to observe the level of effort that the subject is using to

Fig 10.19 Postural strings – front view

The arrows denote the direction of pull of the shoulders.

maintain optimal alignment. If observed (and reported) feedback suggests excessive effort, the practitioner should consider the extent to which this is attributed to muscle shortness and tightness. Where necessary muscle shortness can be avoided with appropriate stretching techniques and tightness can be resolved with muscle release techniques.

Lesson 17: Learning how to walk freely

Theme(s)

Crawling patterns are used to build a progressive foundation for walking. Awareness is also drawn to the use of the eyes as a potential initiator of movement; further explorations are made into intelligent use of the body's centre of gravity and low moment of inertia. The movements in this sequence are used for the purpose of increasing awareness of movement, and should not be regarded as exercises as such. With this in mind, a few purposeful and awareness-rich repetitions of each will provide enough stimuli for learning: perform the sequence in the order shown for maximum benefit.

Postural benefits

Increased efficiency of movement in walking.

Movement sequence

- Neck release (assisted)
- Hip flexor release (assisted)
- Abdominal relaxer (assisted)
- Rocking
- Crawling
- Walking

1. Neck release (assisted)
Refer to lesson 11, step 4.

2. Hip flexor release (assisted)
Refer to lesson 4, step 3.

3. Abdominal relaxer (assisted)

The subject lies supine. The practitioner kneels in front of them and takes hold of both legs. The practitioner slowly pushes the knees towards the subject's chest, until they are positioned over their abdomen. This position is held for 60 seconds, as the subject breathes deeply and freely. This is a passive movement and all effort of holding the legs in this position is provided by the practitioner.

Fig 10.20 Abdominal relaxer

4. Rocking

Begin in a four-point kneeling position with equal weight between arms and legs. Hold this position for a few moments allowing your spine to find a relaxed alignment. Begin by slowly shifting your weight forwards until you are comfortably holding more weight through your arms, and your head is further forwards of your hands. Notice how much effort you are using. Hold this position for 10 seconds, relax, and repeat several times.

Now, in the same position, imagine there is something you really desire about a 0.9m in front of you. If you want, you can place a suitable object in this position. Now, as you begin to lean forwards, initiate the movement by seeking out the object with your eyes – almost as though your eyes are projecting outwards telescopically towards the object. Notice how your head follows and how this movement begins to pull your trunk forwards over your hands. Keep looking at the object and hold this position. How does the movement feel now? How much effort are you using? Are you able to rock further forwards?

You may notice that your spine has extended, which is due to the slight neck extension in reaching your head forwards. Practise this movement several times using as little effort as possible.

5. Crawling

Start in a balanced and relaxed four-point kneeling position as before. Imagine you are about to crawl forwards leading with your right hand. As you think about doing this, you will notice that the weight comes off your right hand. Do this, but leave your right hand on the floor for the moment.

Notice where your body-weight shifts. In the first instance you will probably feel the weight shift over to your left hand. While this is true, where else does a transfer of weight occur? Notice

Fig 10.21 Rocking – eyes down

Fig 10.22 Rocking eyes forward

Fig 10.23 Crawling

that the weight also shifts to your right knee. In doing so, your left hip moves into passive flexion – try to get a sense of this shift. The left hip flexors are almost completely relaxed, so your left thigh hangs like a pendulum. At this point your left knee is still in contact with the ground but is ready to be moved forward with the right hand.

At this point, most individuals will activate the left hip flexors to lift the left leg off the floor, in order to move it forward. However, this is inefficient and costly in energy. The forward momentum of the entire body is actually propagated by the eyes and head, as you discovered in step 4. If you think about projecting the eyes to a point about 0.9m in front of you, while simultaneously reaching forwards with your right hand, you will notice that your entire trunk is pulled along effortlessly. In the process, your left knee moves passively forwards, lightly dragging on the floor; there is no need for active hip flexion to pull the knee forwards. Practise this movement slowly forwards and backwards several times, before swapping sides. You may even progress to making a few crawling movements to reinforce this further.

If you can, watch a baby or toddler crawling, notice how the trail leg simply drags on the floor, and the effortlessness of the entire movement. Notice how they position their head and how they use their eyes. This will help to consolidate your learning further. You are now ready to use this mechanism in walking.

6. Walking

Stand still with good posture for a few moments, sensing how you feel. Take a few steps forward and notice where the movement begins. How are you initiating the movement? In which direction do you move first? How much muscle effort is required? Many will notice that the first movement is sideways, to take the weight off one leg and place it on the other one. The leg is then free to be lifted and placed forward.

Now stand still again and explore a movement or simply rock forwards slightly a few times. Notice how easy it is to fall out of balance by such a small rocking movement. Because the body has such a high centre of gravity, it also possesses a low moment of inertia, so it takes very little energy to move us away from a position of balance. We shall now look at ways to use this information to move forwards.

Fig 10.24 Child crawling

Notice how the left knee is sliding forwards, as opposed to being lifted.

As you stand, push your right knee forwards slightly – notice how it passively flexes, allowing your hip to also flex passively. You may now notice that your right heel comes off the floor slightly. Do this several times using minimal effort. If you rock forwards slightly now, you will notice how your right knee and hip flex more easily, allowing your trunk to move forwards over your legs freely. You are now learning how to make use of almost passive motion at the hip and knee to create forward momentum. The final step is to do exactly that: take a step.

As you lean forwards and flex your knee and hip, allow the forward momentum to carry you further – at some point you will have to step forwards with your right foot and continue the movement. Notice that there is no need to shift weight sideways unnecessarily, nor is their any need to use excessive amounts of active hip flexion – you simply fall into the movement and continue it with the other leg.

Don't be surprised if this action feels a little strange and disjointed at first – you are re-wiring many years of walking in a different way. You are now learning that walking is just a series of controlled falls in a forward direction – all you have to do is step into the falls in a coordinated manner. Aim simply to move your knees forward rather than unnecessarily overusing your hip muscles. For inspiration, look no further than babies learning to walk and you clearly see the forward falling action (albeit less coordinated) that they use to walk. For a full postural workout, practise walking with postural strings, and begin to notice how effortless walking can actually be.

Fig 10.25 Walking – passive flexion of hip and knee

Fig 10.26 Child walking

Note the slight forward lean into the move ment allowing almost passive hip flexion.

REFERENCES

Alexander FM (1932). *The use of the self: its conscious direction in relation to diagnosis, functioning and the control of reaction.* London: Chaterson.

Aston J (1999). *Aston Postural Assessment Workbook: Skills for Observing and Evaluating Body Patterns.* Psychological Corp.

Belenkiy VY, Gurfinkel VS, Pal'tsev YI (1967). Elements of control of voluntary movements. *Biofizika* **10**: 135–141.

Chaitow L, DeLany J (2000). *Clinical applications of neuromuscular technique - Volume 2: The lower body.* Edinburgh; New York: Churchill Livingstone.

Chek P (2001). *Movement that matters.* San Diego: C.H.E.K. Institute.

Feldenkrais M (1949). *Body and Mature Behaviour. A study of anxiety, sex, gravitation & learning.* Berkley, California: Frog Ltd and Somatic Resources.

Feldenkrais M (1981). *Elusive Obvious or Basic Feldenkrais.* Cupertino Ca: Neta Publications.

Feldenkrais M (1985). *The potent self: a study of spontaneity and compulsion.* Berkley, California: Frog Ltd and Somatic Resources.

Gowitzke BA, Milner M (1988). *Understanding the scientific bases of human movement.* 3rd ed. Baltimore: Williams and Wilkins.

Hanna T (1988). *Somatics: reawakening the mind's control of movement, flexibility, and health.* Cambridge, Massachusetts: Perseus Books.

Hayes KC (1982). Biomechanics of postural control. *Exerc Sport Sci Rev* **10**: 363–391.

Horak FB, Nashner LM (1986). Central programming of postural movements: adaptation to altered support-surface configurations. *J Neurophysiol* **55**(6): 1369–1381.

Janda V (1994). Chapter 10: Muscles and motor control in cervicogenic disorders: assessment and management. In *Physical therapy of the cervical and thoracic spine.* Grant R (eds.). 2nd ed. New York; Edinburgh: Churchill Livingstone. pp ix, 442p.

Jull G, Janda V (1987). Muscles and Motor Control in Lower Back Pain–Assessment and Management. In *Physical therapy of the low back.* Twomey LT and Taylor JR (eds.). New York: Churchill Livingstone. pp 328.

Kendall FP, McCreary EK, Provance PG (2005). *Muscles, testing and function: with posture and pain.* 4th ed. Baltimore; London: Williams & Wilkins.

Massion J (1992). Movement, posture and equilibrium: interaction and coordination. *Prog Neurobiol* **38**(1): 35–56.

McGill S (2004). *Ultimate Back Fitness and Performance.* Stuart McGill.

Myers TWLMT (2001). *Anatomy trains: myofascial meridians for manual and movement therapists.* Edinburgh: Churchill Livingstone.

Nashner LM, Cordo PJ (1981). Relation of automatic postural responses and reaction-time voluntary movements of human leg muscles. *Exp Brain Res* **43**(3–4): 395–405.

Norris C (2000). The muscle debate. *Journal of Bodywork and Movement Therapies* **4**(4): 232-235.

Park S, Toole T, Lee S (1999). Functional roles of the proprioceptive system in the control of goal-directed movement. *Percept Mot Skills* **88**(2): 631–647.

Rolf IP (1989). *Rolfing: reestablishing the natural alignment and structural integraton of the human body for vitality and well-being.* Rochester, Vt.: Healing Arts Press.

Sahrmann S (2002). *Diagnosis and treatment of movement impairment syndromes.* St. Louis, Mo; London: Mosby.

Todd ME (1937). *The Thinking Body. A study of the balancing forces of dynamic Man, etc. [With illustrations.].* pp. xxiv. 314. Paul B. Hoeber: New York, London.

Winter DA, Prince F, Frank JS, et al. (1996). Unified theory regarding A/P and M/L balance in quiet stance. *J Neurophysiol* **75**(6): 2334–2343.

SUGGESTED READING

Alexander, FM. *The Alexander Technique: The Essential Writings of F. Matthias Alexander.* University Books, New York, 1989.

Barlow, Wilfred. *The Alexander Principle: How to use your body without stress.* Orion Books Ltd, 2001.

Chaitow, Leon and DeLany, Judith Walker. Clinical Application of Neuromuscular Techniques. Volume 2. *The Lower Body.* London: Churchill Livingstone; 2002.

Chek, Paul. *How to eat, move and be healthy.* CHEK Institute, San Diego, 2004.

Cook, Gray. *Athletic Body in Balance.* Human Kinetics, 2003.

Corning-Creager, Caroline. *Therapeutic Exercises using Foam Rollers.* Executive Physical Therapy, Inc. 1996.

Feldenkrais M. *Awareness Through Movement.* Harper and Row. New York, 1972.

Feldenkrais, M. *Body & Mature Behavior: A study of anxiety, sex, gravitation & learning.* North Atlantic Books. 1949.

Feldenkrais, M. *Mindful Spontaneity: Lessons in the Feldenkrais Method.* North Atlantic Books. 1996.

Feldenkrais, M. *The Master Moves.* Meta Publications. 1984.

Franklin, Eric. *Dynamic Alignment Through Imagery.* Champaign, IL: Human Kinetics. 1996.

Franklin, Eric. *Dynamic Alignment Through Imagery.* Human Kinetics Europe Ltd. 1996.

Hamilton, Nancy and Luttgens, Kathryn. Kinesiology. *Scientific Basis of Human Movement.* International Edition. McGraw Hill; 2002.

Hanna Thomas L. *Somatics: Reawakening the Mind's control of Movement, Flexibility and Health.* Reading, MA. Perseus Books, 1988.

Hanna Thomas L. *The Body of Life: Creating New Pathways for Sensory Awareness and Fluid Movement.* Rochester, Vermont. Healing Arts Press 1993.

Hanna, Thomas. *The Body of Life – creating new pathways for sensory awareness and fluid movement.* Healing Arts Press. 1993.

Kendall, Florence, McCreary, Elizabeth and Provance, Patricia. *Muscles Testing and Function, Fourth Edition with Posture and Pain.* Philadelphia, Pennsylvania. Lippincott Williams and Wilkins; 1993.

Langford, Elizabeth. *Mind and Muscle: An owners handbook.* Garant, 2001.

McGill, Stuart. *Low Back Disorders. Evidence-Based Prevention and Rehabilitation.* Champaign, IL; Human Kinetics; 2002.

National Academy of Sports Medicine. *Integrated Strength Training Concepts.* Certified Personal Trainer Course Manual, 2002.

Norris, Christopher M. Back Stability. Champaign, IL; Human Kinetics; 2000.

Norris, Christopher M. *The Complete Guide to Stretching.* 2nd Edition, London. A & C Black, 2004.

Patel, Kesh. *Corrective Exercise: A Practical Approach.* Hodder and Stoughton. 2005.

Richardson, Carolyn et al. *Therapeutic Exercise for Spinal Segmental Stabilization in the Low Back.* Scientific Basis and Clinical Approach. London: Churchill Livingstone; 2000.

Sahramann, Shirley. *Diagnosis and Treatment of Movement Impairment Syndromes.* St Louis, Missouri. Mosby Inc; 2002.

Zemach-Bersin, Kaethe, Zemach-Bersin, David, and Reese, Mark. *Relaxercise: The Easy New Way to Health and Fitness.* Harper, San Francisco, 1989.

GLOSSARY

Adaptive shortening – The process by which a muscle structurally shortens as an adaptation of being held in a prolonged passive position; muscle shortness causes misalignment

Alignment – An arrangement of the joints of the body in relation to one another and gravity

Altered dominance – Increased activity of synergist muscles, producing a movement that occurs with altered neurological and mechanical control

Anteversion – A structural misalignment in relation to the femur, where the angle of the head and neck of the femur is rotated anteriorly beyond the normal alignment of 15°

Atrophy – A decrease in muscle size due to inactivity

Balance – The maintenance of the body's centre of gravity over the base of support

Bipedalism – The practice of standing or moving upright on two feet; the method of human locomotion

Cartesian – Relating to the thoughts and methods of the philosopher/mathematician René Descartes

Centre of gravity – The point around which a body's weight is equally balanced

Contract–relax techniques – An umbrella term for a number of techniques that increase passive flexibility of muscles; in its simplest format, a muscle is isometrically contracted against resistance, immediately followed by a passive stretch

Contralateral – Affecting or involving the opposite side of the body

Coordination – The control over a series of muscle contractions, to create a desired motion

Configuration – The way the parts of the body are physically arranged and connected that allow for optimal structure and function

Differential movement – A process that usually involves two movements, where one is performed in opposition to the other

Efficiency – The ability of the body to organise itself without wasted effort or energy

Endurance – The ability of muscles to perform a sub-maximal task without fatigue (just below maximal effort)

Equilibrium – A stable, constant condition that is controlled by interrelated regulation mechanisms; also known as homeostasis

Ergonomic – Designed for maximum comfort, safety and efficiency, especially relating to the workplace

Eversion – Turning of the foot outwards away from the body; a combination of pronation and forefoot abduction

Exteroception – The ability to sense stimuli arising outside the body. The key exteroceptive senses are vision, hearing, taste, smell and touch

Feedback – The return of proprioceptive information to the body in a way that allows it to adjust or modify its performance, usually to maintain equilibrium or some desired state

Force couple – Action of two forces in opposite directions to produce rotation around a joint

Friction – The resistance encountered by an object moving in relation to another object with which it is in contact

Gait – Any pattern of locomotion; a way of walking, running, or moving on foot

Homeostasis – See **Equilibrium**

Hypertonicity – Excessive muscle tone or muscle tension

Hypertrophy – An increase in muscle size; occurs through an increase in the size, rather than the number of muscle cells

Inertia – The property of a body that resists change in its motion

Integration – A systematic and often thematically-organised method of opening up the potential for movement, usually following msucle release and stretching

Inversion – Turning the foot inwards towards the body; a combination of supination and forefoot adduction

Ipsilateral – The same side of the body

Kyphosis – Increased convexity in the curvature of the thoracic spine as it is viewed from the side

Lordosis – Increased concavity in the curvature of the lumbar spine as it is viewed from the side

Mobility – The ability to move freely; often regarded as a combination of flexibility and coordination

Muscle testing – A manual method of testing that assesses the structural and functional characteristics of muscle, such as length, strength, and dominance

Muscle length – The ability of a muscle to confer range of motion around a joint

Muscle strength – The ability of a muscle to confer stability around a joint

Pandiculation – A series of strong muscular contractions resulting in motor cortical arousal, such as yawning or stretching

Paraspinal – The muscles next to the spine that provide support and movement

Perturbation – A disturbance of the postural system by any internal or external influence

Phylogenetic – Relating to the evolutionary history of a particular group of organisms

Plumb line – A vertical line of reference that represents the line of gravity; used in postural assessment

Positional release – A series of techniques that work on the principle of approximating the two ends of attachment of a muscle thereby reducing its effort of contraction

Postural adjustments – Any adjustment or mechanism of adjustment that restores postural homeostasis or equilibrium

Pronation – Rotation of the foot to position the sole laterally

Prone – Lying face downwards

Proprioception – The ability to sense stimuli arising within the body; the body's awareness of position, posture, movement and changes in equilibrium

Prophylactic – Used to describe movements or exercises that are preventative

Q-angle – The angle formed by a line drawn from the anterior superior iliac spine to the centre of the patella, and a second line drawn from the centre of the patella to the tibial tubercle: the Q-angle in adults is approximately 15°

Reciprocal activation – By which muscles co-contract to maintain joint stiffness

Reciprocal inhibition – The mechanism by which a muscle is inhibited by a hypertonic antagonist; it is an essential, automatic neurological function, designed to provide optimal joint function

Retroversion – A structural misalignment in relation to the femur, where the angle of the head and neck of the femur is rotated posteriorly beyond the normal alignment of 15°

Righting reflex – Any reflex that brings the body into normal position in space, and resists forces acting to displace it out of its normal position; a normal reaction is dependent on normal vestibular, visual and proprioceptive functions

Scapulohumeral rhythm – The coordinated movement of the scapula and humerus during arm movements

Scoliosis – A lateral curvature and rotation of the spine, either congenital or acquired

Shear force – A force that can result in deformation of contiguous body parts as they slide in relation to one another in a direction parallel to their plane of contact

Somato-sensory – Referring to the sensory signals from all tissues of the body including skin, viscera, muscles, and joints

Stability – The capacity to provide support

Strain – The extent of deformation of a tissue under loading

Stretch reflex – Reflex contraction of a muscle when an attached tendon is pulled; important in maintaining erect posture

Supine – Lying face upwards

Supination – Rotation of the foot to position the sole medially

Tension – The effective force generated by a muscle

Torsion – The stress or strain placed on a body part that has been twisted

Torque – Force which tends to cause rotation

Unilateral – Affecting or involving only one side of the body

Vestibular system – The system in the body that is responsible for maintaining the body's orientation in space, balance, and posture; also regulates locomotion and other movements

Windlass Effect – The mechanism by which the plantar fascia is tensioned as the toes dorsiflex during the toe-off phase of gait, resulting in a lifting of the arch; this mechanism helps to create a rigid lever for efficient forward propulsion

INDEX

Note: Some lesson plans include references to exercises described elsewhere. Page numbers for these references are in brackets.

ALSO AVAILABLE

The Complete Guide to Stretching – 3rd edition
Christopher M. Norris

A reasonable level of flexibility is essential to the healthy functioning of joints and muscles, which in turn facilitates performance and reduces the risk of injury. *The Complete Guide to Stretching* provides an accessible overview of the scientific principles that underpin this form of training and offers more than seventy exercises designed to safely increase range of motion right across the body.

The Complete Guide to Stretching is the definitive practical handbook for:
- sports participants and recreational exercisers who are keen to achieve a level of flexibility that will enhance their performance
- sports coaches and fitness instructors who are seeking a thorough understanding of the principles and practice of this often neglected component of physical fitness
- sport and exercise therapists who could use stretching as an important part of a balanced rehabilitation programme.

This new edition is in full colour, with brand new photographs demonstrating the stretches throughout.